An Introduction to

ENGINEERING
and
ENGINEERING DESIGN

SECOND EDITION

E. V. KRICK

Lafayette College
Easton, Pennsylvania

P9-BBP-618

John Wiley & Sons, Inc.

NEW YORK · LONDON · SYDNEY · TORONTO

Copyright © 1965, 1969 by John Wiley & Sons, Inc.
All rights reserved. No part of this book may
be reproduced by any means, nor transmitted,
nor translated into a machine language without the written permission of the publisher.

10 9 8 7 6 5 4 3 2

Library of Congress Catalog Card Number: 68-8106
Cloth: SBN 471 50740 7   Paper: SBN 471 50741 5
Printed in the United States of America

If you are considering this book for a course, by all means ask for a copy of the teacher's manual. The manual includes outlines, objectives, problems, and other aids to an introduction to engineering. It is especially important if you are interested in design problems. I much prefer to present a problem in a manner that approximates real life, instead of telling the students to "work problem 2 at the end of the chapter." Therefore a number of problems that would otherwise be in the book are in the manual instead, so that you can introduce them in a realistic and motivating manner. These are freshman-level design problems in the form of letters, memoranda, and the like, which you can reproduce and distribute.

In Chapter 1 I have outlined the major objectives of this book. Here are some comments on these objectives that may interest you as a teacher. One objective is to introduce engineering to a young person who is considering or beginning an education in this field. There are at least five ways of doing this, one or more of which will surely be meaningful to the student:

1. Presenting case studies of engineering in action.
2. Describing the origin and nature of contemporary engineering.
3. Outlining the important attributes of an engineer.
4. Describing the process of design. By "design" I mean the sequence of activities that starts with recognition of a problem and ends when a functional, economical, and otherwise satisfactory solution to that problem has been specified. It includes problem definition, analysis, synthesis, invention, performance prediction, decision making, optimization, specification, and in fact most skills and techniques considered part of the engineering method. Describe the design process and you describe the essence of engineering.
5. Involving the student in design.

This book does the first four; an accompanying course should do the fifth.

Another objective is to *start* developing the student's skill in modeling, computer *application,* optimization, design, and selecting the best approach to each problem he faces. I have elaborated on these skills primarily because they are basic and in my opinion are not satisfactorily described at an introductory level in engineering literature. Satisfactory books on slide-rule use, measurement, computer programming, experimentation, graphics, and other engineering skills are available. Not so for the skills I have introduced in detail. This book provides a foundation from which the development of these skills can proceed.

Still another objective is to improve the student's understanding of the goals of an engineering education. It is especially important but difficult for him to appreciate the ultimate purpose and significance of courses he takes in the first two years. Many students do not see the importance of ability in verbal and graphic communication. And they often ask, "Why must we take so many nontechnical courses?" Then, too, their notions of the manner in which engineers use mathematics are vague. Their motivation suffers when they fail to see the relevance of these courses to engineering practice. Therefore I have explained how these subjects relate to an engineer's work. If satisfactory reasons exist for each phase of the engineer's formal training, then by all means they should be made known to the student! This book should help carry the student through the motivationally trying first two years.

E. V. KRICK

CONTENTS

AN INTRODUCTION TO ENGINEERING
AND ENGINEERING DESIGN

OBJECTIVES

What is engineering? What is an engineer's day-to-day work like? What abilities are important to success and satisfaction in this field? What are some of the problem areas that will provide exceptional opportunities and challenges for engineers? What are some of the major benefits to be had from an engineering education? If you seek answers to these questions, this book is addressed to you.

If you are a precollege student in the process of investigating alternative careers, you should be asking such questions and obtaining *factual answers.* I emphasize factual answers because of the prevalence of misconceptions. The news media and most people seem confused about the comparative roles of scientists and engineers. Also, many still hold the outmoded "leather boots and transit" and "drawing board" images of engineers. Of course misconceptions and vague notions are expected because the public rarely sees the engineer at his job. We get to observe the teacher and the doctor in action, but not the engineer. No wonder so much advice and so many popular generalizations about engineering are erroneous. Yet it is vital that you obtain reliable information; an unwise career choice can be costly. Investigate thoroughly, consider your sources, become well informed before you select a career from the many alternatives.

If you are studying engineering, you too should be seeking answers to the above questions, especially if you are in the first two years of the curriculum. Most of the courses in this period deal with the basics of chemistry, physics, mathematics, and similar subjects. Essential as they are, they are poor indicators of what engineering practice is like, so don't make the mistake of basing your career decision on the courses in these first two years. Yet especially in this period there must be questions that are bothering you, including how certain of your required courses will ever be useful, whether you really have what it takes to be a successful engineer, and what avenue of specialization you should follow.

Therefore this book was written to . . .

- give you a realistic picture of engineering—what it is, requires, and offers, and the day-to-day activities it involves;
- help you decide whether this is the career for you;
- start your development of certain skills frequently employed by engineers—frequently enough, in fact, that some familiarity with them tells you a lot about the practice of engineering;
- clarify the objectives of an engineering education, including the purposes of different types of courses;
- familiarize you with some terms widely used in engineering literature and conversation;
- arouse your interest in effectively solving some of the pressing problems confronting mankind.

I hope this book achieves these objectives for you. Of course, you must also do your part.

ENGINEERING PROBLEMS

WHAT constitutes a problem? You use this term often; what does it mean? What do all problems have in common? I suspect that your answers to these questions are vague. Yet a discussion of engineering is essentially a discussion of problems and problem solving. Therefore you should acquire accurate definitions of these terms now. And so this book begins with an introduction to the general characteristics of problems.

A problem arises from the desire to achieve a transformation from one state of affairs to another. These states might be two locations the interval between which must be traversed. The problem may be to get from one river bank to the opposite one, from one city to another, from one planet to another. Other problems involve a transformation from one form or condition to another, for example, bread to toast. In any problem there is an originating state of affairs; call it "state A." Similarly, there is a state of affairs the problem solver seeks a means of achieving; call it "state B." Note that this is true of personal problems, communication problems, business problems, and in fact *all* problems (Figure 1).

A solution is a means of achieving the desired transformation. A problem to which there is only one possible solution is rare indeed; in most problems there are many alternative solutions, many more than there is time to investigate. Think of the many modes of travel and all the possible routes they can be combined with to provide alternative means of getting from one point on earth to another.

In addition, a problem involves more than finding *a* solution; it requires finding a *preferred* means of achieving the desired transformation; for example, the mode of transportation that is best with respect to cost, speed,

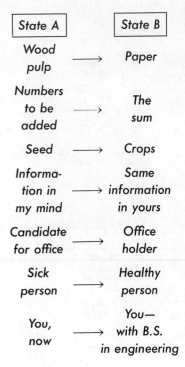

Figure 1

State A	State B
Wood pulp	Paper
Numbers to be added	The sum
Seed	Crops
Information in my mind	Same information in yours
Candidate for office	Office holder
Sick person	Healthy person
You, now	You— with B.S. in engineering

safety, comfort, and reliability. A basis of preference among various solutions is called a *criterion*.

Finally, it is difficult to imagine a problem in which there are no restrictions on solutions. A restriction is something that *must* be true of a solution. Examples: a high school student has decided that the college he attends *must* be coeducational; certain characteristics of structures are fixed by building codes; light, water, and nutrients *must* be provided to transform a seed into a plant.

Restrictions, criteria, alternatives, and the dominant characteristic of any problem—a transformation—will be conspicuous in the following descriptions of engineering projects. I trust you will become conscious of them, for as an engineer you should be skilled at identifying the basic characteristics of the problems you are to solve.

Engineering in Action

An Information-Processing System The management of Computer Electronics Company believes that there is a promising market for an information-processing device conceived by one of its engineers. This machine, which he refers to as the Diagnosticator, will assist in the diagnosis of human illnesses in the following manner. The physician examines a patient as he has always done, and then communicates his findings to the Diagnosticator. The machine processes this information and returns a list of ailments and for each ailment the probability (the chances) that it is the patient's trouble. For example, in response to a specific set of symptoms and certain information about the particular patient (e.g., age, weight, and habits), the device says that there are 63 chances out of 100 that the patient has ailment A, 18 out of 100 that his illness is B, and so on.

Before this company will manufacture such a device, the management must be convinced that a worthwhile profit can be realized from the venture. Therefore the engineer who originated the idea has been commissioned to provide preliminary specifications for this device *and* a forecast of the costs of completing its development and of manufacturing it. If after this "first approximation" it appears that the machine can be developed and manufactured at a cost that will ultimately yield an attractive return

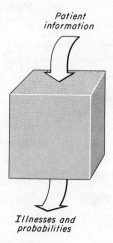

Patient information

Illnesses and probabilities

on the investment, he will then complete his design in all necessary detail.

Among other things, he has been told that this device must produce the desired diagnostic information within one minute, must not be cumbersome, and must operate on the ordinary household power supply. The management wants these preliminary specifications and the predictions of costs within two months.

Commentary. He views this problem as that of finding the most effective means of transforming patient information (state A) into a set of ailment probability statements (state B). The device he designs must satisfy the restrictions (e.g., must produce results within one minute). In addition, his solution must be the best with respect to manufacturing cost (a criterion) and appeal to potential buyers (another criterion) that he can devise in the two months allotted for the project. This in brief defines his problem.

During the project this engineer uses his knowledge and inventiveness to devise a variety of possible solutions. One consists of a number of typewriter-size electronic devices located in user's offices (margin). With these devices doctors or their nurses transmit symptom information to, and receive results from, a centrally located information-processing unit that serves a number of customers (Figure 2). An alternative to this plan is to provide each user with an independent unit that does the whole job in his office. Some relative merits of these two alternatives are obvious, but what these mean in dollars is not so apparent.

He is also investigating alternative methods of entering data into and getting results out of these machines, alternative ways of processing patient data to obtain the desired results, and many types of components. These and numerous other possibilities in a variety of combinations yield a large number of alternative systems, all of which are feasible but not equally desirable. He must evaluate these competing schemes and reduce them to the one best system.

Throughout the project he works with a number of persons in a variety of capacities. Among them are marketing specialists from whom he learns the typical potential user's preferences concerning various features of the product. He also consults medical diagnosticians. He works closely with manufacturing experts in estimating what it will cost to manufacture a given version of his device. He works

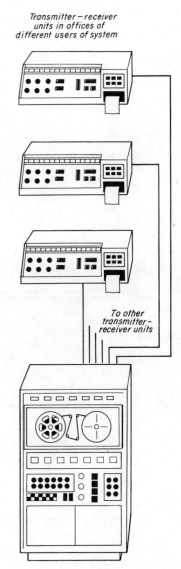

Transmitter – receiver units in offices of different users of system

To other transmitter – receiver units

Central processor, services many transmitter – receiver units

Figure 2   The engineer had an artist prepare sketches of the major alternatives he is considering, for inclusion in his report to the management. This one is of the centralized Diagnosticator system.

closely with his firm's executives, with telephone people concerning use of their lines for transmission of information, and with numerous other persons. He has long since learned that his work involves more contact with people than he realized before his first engineering job. He has also concluded that, in general, people problems are more frustrating than technical ones.

The Diagnosticator is a challenging assignment, one reason being that such a system is not yet marketed, so that the effort is a pioneering one. Furthermore the project is this engineer's baby from the start; he receives no direct supervision, and what he does during the two months is up to him—but results *are* expected. This is the kind of opportunity you will like, especially when your creation, like this one, stands to be of special benefit to the public. But don't expect this freedom and responsibility until you have had experience—this engineer has nine years of it. You will start your employment career working on relatively small parts of projects, under close supervision. Freedom and responsibility will come with experience. This is fortunate, for if you were thrown into an assignment like this one in the first few years of employment you would flounder.

At the conclusion of his investigation the engineer will present his recommendations through an extensive written report and an oral briefing to top executives. He will include forecasts of the costs associated with the manufacture of his design. The importance of this information to the future of the project is obvious.

An Automatic Production Machine   In the Bell Telephone System there are millions of switches in operation. A long and persistent engineering effort has made them able to function for years and to make billions of connections without breaking down. But they are still subject to failure caused by moisture or foreign matter. The cost of isolating and remedying such failures, plus the cost of trying to prevent trouble through elaborate maintenance and cleanliness measures, have been a matter of concern to the company for years. An engineer for the company was asked to recommend a means of reducing these costs and improving the reliability of the system.

In the course of his investigations this engineer developed and evaluated many possibilities, finally selecting the rather ingenious type of switch pictured in Figure 3 as the most promising solution. It consists of two flat metal elements called reeds, which are encased in a gas-filled, sealed, glass tube. Shown here are the two identical reeds, the glass tube in which they are to be encased, the reeds sealed in a gas-filled glass envelope, and a close-up view of the contact point. In use the switch is mounted in a small electromagnetic coil that when energized causes the reeds to make contact and complete an electric circuit. This device is remarkably fast, very reliable, maintenance-free, and in most respects superior to any other switch yet conceived. However, a very important question—one that determines whether this novel switch can ever be of much use to the company and its customers—remains to be answered. Can it be economically manufactured by the millions?

To answer this question, a team of engineers was assigned the task of developing, if feasible, an economical method of making these switches. The solution is the remarkable machine shown in Figure 4.

Commentary. This team's problem was to find the most economical means of transforming glass tubing, reeds, and gas into the specified switches—by the millions. It is a challenge to develop a machine that will place the reeds in a glass tube and align them to the close tolerances required. Yet if these millions of switches were to be made by hand, a small army of people would be required, at prohibitive cost. Hence the usefulness of the switch depended on the team's ability to develop an economical machine.

As in any engineering endeavor, during this project the matter of economy of the venture was under constant scrutiny. Periodically the team paused to re-evaluate the probability of its being able to develop an economical machine. If at the outset or at any time during the project it had appeared that any method of manufacture that could be developed would be prohibitively costly, the originator of this switch would have been back looking for another solution to the problem—one cheaper to manufacture.

The designer of the switch specializes in the development of devices employed in the telephone system. The engineers subsequently assigned to the project develop means of manufacturing these devices. They are usually referred to as process or manufacturing engineers. Members of design teams like this are usually complementing experts. In this particular case one specializes in the be-

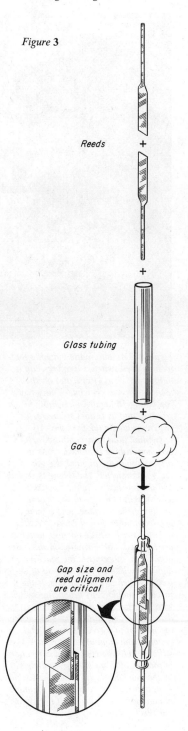

Figure 3

Reeds +

+

Glass tubing

+

Gas

Gap size and reed aligment are critical

Figure 4   *An automatic machine for making reed switches at more than a million per year. It operates on the merry-go-round principle. The turret containing 18 identical assembly heads revolves, and as it does so the reed switch gradually takes shape. At sequential stations around the turret's periphery tubing is placed, reeds are inserted, they are aligned, the gap is set, gas is injected, the tubing is sealed, and the finished switch is ejected. The switches then go to the testing section of the machine, where their physical and electrical characteristics are measured. Unsatisfactory switches are rejected. Furthermore, on the basis of these measurements the machine adjusts itself to correct a cause of defective switches. For example, if the machine should start producing switches with oversize gaps between reeds, it detects this and adjusts the gap-setting mechanism to correct the error and restore production of acceptable switches. This detection and correction of sources of unsatisfactory switches by the machine itself, without assistance from a human operator, is an example of automation. (Courtesy of the Western Electric Company.)*

havior and forming of glass, another in the behavior of machine components and mechanisms, another in electrical and magnetic phenomena, and so on. A close working relationship between members of an engineering team is vital; there must be a great deal of interaction between the knowledge, ideas, and decisions of the various specialists concentrating on different facets of a problem. One member of this group, commonly referred to as the project (or systems) engineer, served primarily as a co-ordinator of the activities of the others, in order to ensure that all parts of the final system would be properly interrelated.

When the group believed that it had devised the most economical machine, their proposal had to be specified in complete detail so that technicians and craftsmen could construct a prototype version. The engineers were responsible for overseeing the construction of this prototype. They found it necessary to make some modifications of their original design during this construction period. When the model was complete, they supervised test runs of the machine. Additional design modifications were instituted as a result of these trial production runs. Finally, after a lengthy period of testing and refining, the proposed machine was considered ready. Complete specifications of the

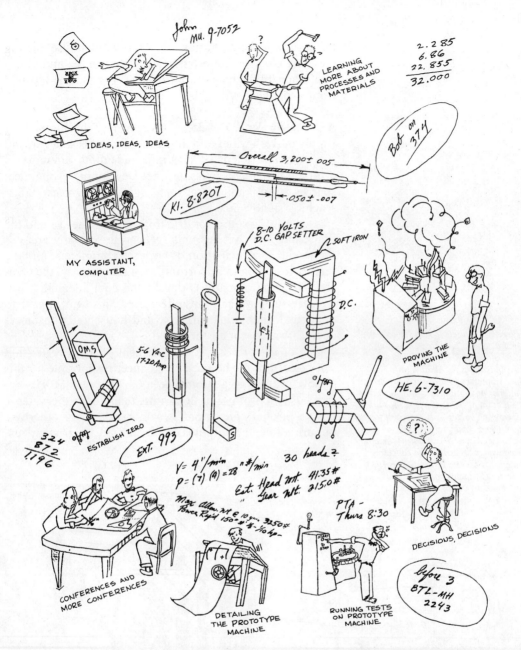

prototype model as it finally evolved were prepared by draftsmen so that additional machines could be constructed. As a result the more effective switch became available for general use at a rate of many millions per year.

Even then the task was not complete. The engineers fol-

Figure 5   A page from the notes of one of the team's engineers reflects by way of his doodlings the many alternatives, false starts, meetings, details, procedural steps, and communications involved in the development of the machine shown in Figure 4. (Courtesy of the Western Electric Company.)

lowed up on their creation by observing it in use, recommending changes in design where appropriate, and evaluating their brainchild so that future projects might benefit from experience with this machine.

A Domestic Water Desalter   As a result of the diminishing supply of fresh water and of the rapidly increasing demand for it, the problem of providing adequate amounts of drinking water is a pressing one. Developing economical sources of drinkable water is *an engineering problem* of far-reaching significance.

A promising source of drinking water is the sea and the brackish water that lies underground in so many areas of the world. In anticipation of both the commercial and the service-to-mankind opportunities in this activity, the General Electric Company is developing equipment that will convert such water to usable form. This will appeal to municipal, industrial, military, and household consumers of fresh water around the world.

One of the projects is development of a converter that can be used in the home. The engineer responsible for this is now evaluating a prototype model (Figure 6). His design has real promise; it does more than desalt the water—it also purifies. The input is salt, brackish, or otherwise impure water; the output is demineralized and pasteurized.

Figure 6 ` Household unit for converting brackish or sea water. Close-up view shows the actual conversion mechanism. As the shaft rotates, the discs connected to it are coated with a thin film of warm brackish water in the bottom of the tank. This thin film of water vaporizes as it passes through the air and condenses on the cool stationary plates. This condensate, now fresh water, drops into the collecting troughs. (Courtesy of the General Electric Company.)

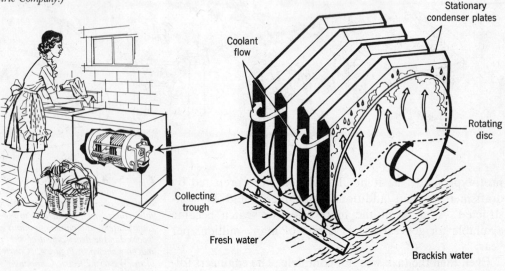

Therefore this converter should be useful in homes, small commercial establishments, military field units, and on shipboard. It is effective and simple, needs very little maintenance, and does not require boiling water or a pressure vessel.

Commentary. Finding *a* method of converting sea and brackish water to fresh water is not the problem; distillation has been known for centuries. The problem is to find a means of transforming *quantities* of such water to usable form *at a cost a substantial number of potential buyers will accept.*

The development of this converter is based partly on the engineer's background of technical and scientific knowledge (yes, he *was* applying things he learned in college physics and chemistry) and partly on his inventiveness. He could not have developed this machine without understanding vaporization and condensation phenomena, the behavior of thin films of liquid, thermal processes, and other scientific facts. However, this knowledge alone would never yield the particular device that evolved. The idea of the rotating discs interspersed with stationary collector plates, the particular configuration of these plates, and other unique features of this mechanism are products of the process called *invention.* You won't find such things in handbooks or textbooks; they are fruits of the mind's creative powers.

Some real engineering effort and talent have gone into this machine. Many different conversion schemes were evaluated; hours and hours of tests were made; considerable research was necessary. The result of this extensive development process is a well-engineered device that will prove successful as a financial venture and valuable as a service to the public.

Why was this intensive engineering effort necessary? The device looks so simple. And it is, but this simplicity is deceiving. It leads you to under-estimate the effort, ingenuity, analytical work, and investigation that went into this device. If all of this had *not* been put into it, the result would probably be more complicated and therefore more impressive to the untrained eye—but no more effective. In fact the more complicated version would be more liable to break down, more costly to make, and perhaps too expensive to sell.

The Chesapeake Bay Bridge-Tunnel A remarkable engineering achievement is the 18-mile structure spanning the mouth of Chesapeake Bay (Figure 7). It is the longest man-made fixed crossing of navigable ocean. This 140-million-dollar structure is a combination of trestles, bridges, and tunnels that carry a heavy flow of vehicular traffic and withstand ocean waves and rushing tides.

Commentary. As is often the case, this span was designed by an engineering consulting firm that makes a business of designing such structures. This firm was commissioned to locate, design, and supervise construction of the entire structure. An unusual restriction was imposed on the nature of this structure, namely, that it could not pass over the main shipping channels because a bridge above them could be bombed and thus trap Navy vessels in the bay. Therefore it was necessary to go under the channels, employing two mile-long tunnels where bridges would have been placed under ordinary circumstances.

A point worth emphasizing is the careful attention that engineers must give to the means of constructing their creations. In fact, especially in a case like this, the methods of construction are as much a part of the problem as the design of the structure itself. After examining Figures 8-12 in succession you will see what I mean.

Trestle piles, cross pieces, roadway slabs, tunnel sections, and other components were prefabricated by mass production methods on land, where construction is achieved with less difficulty and expense. The use of prefabricated components is a method of construction that obviously affects the physical characteristics of the structure itself. Here, as in most engineering problems, there is a strong interdependence between a structure's physical characteristics and the means of constructing it; each significantly affects the other. Furthermore, the design of special construction equipment such as the Two-Headed Monster was a major part of the problem. In this project the cost of equipment and labor required to prepare, transport, and place the stone, concrete, sand, and steel involved many millions of dollars. (The investment in construction equipment employed on this job was about 15 million dollars.) Thus the economic feasibility of a venture like this depends heavily on the engineers' ability to design a structure that minimizes construction cost while satisfying all functional requirements, *and* to devise economical methods of putting it together.

13

Figure 7   The view of the Chesapeake Bay bridge tunnel shown on page 13 looks south from Cape Charles toward Virginia Beach. The Atlantic Ocean is at the left, the bay at the right. The view above shows a portion of the more than 12 miles of trestle employed in this crossing. (Bridge-tunnel photographs courtesy of Chesapeake Bay Bridge-Tunnel District.)

Visible here are the major components of this structure: concrete piles, concrete cross beams, and concrete slabs that form the roadway.

Starting from Virginia Beach, a vehicle moves over three miles of concrete trestle and then, at the first of four man-made islands, enters a one-mile tunnel that carries traffic beneath a major shipping lane. The vehicle continues over four more miles of trestle, through another tunnel, then over more trestle, two bridges, and a natural island before it reaches the mainland at Cape Charles.

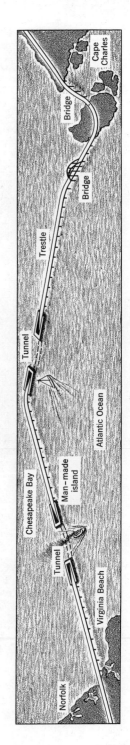

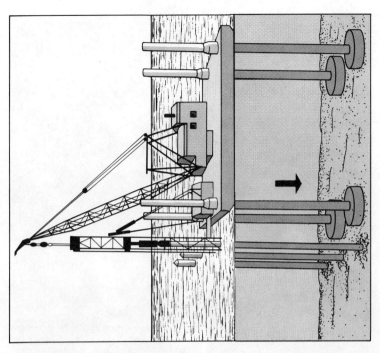

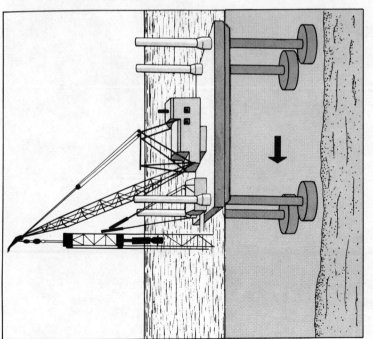

Figure 8 The twelve miles of trestle were constructed by driving long (up to 172 feet) concrete piles deep into the bed of the bay by means of a specially designed machine nicknamed Big D, shown on the opposite page. This 1.5-million-dollar rig is an enormous pile driver and crane mounted on a barge with 100-foot-long, 6-foot-diameter retractable legs. Big D floats into position and then becomes a stationary platform by extending its legs and hydraulically lifting itself out of the water. It "stands" in this position until its work is complete at that spot. Without this stability Big D would never be able to locate the huge piles to close tolerances.

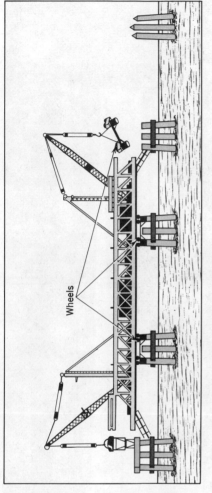

Wheels

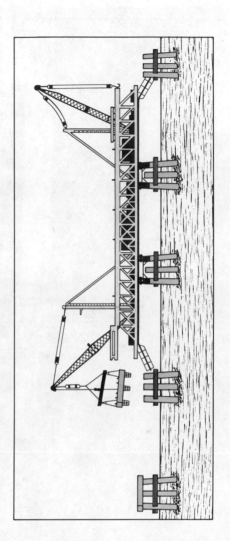

Figure 9   Big D is followed by another special rig called the Two-Headed Monster, shown above. As it crawls along the tops of the piles, the leading "head" of the Monster trims the piles to a uniform height. Cross beams are placed by the Monster's trailing "head."

The Monster rolls along on wheels that are temporarily mounted on the tops of the piles. It has an extra set of wheels which it places on the next row of piles ahead of it, as shown at right.

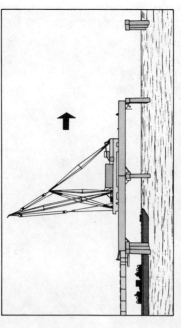

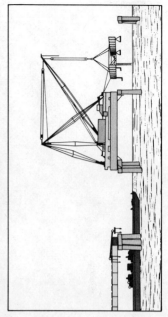

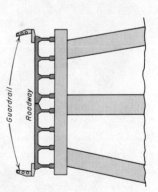

Figure 10   Then comes this machine, called the Slabsetter, which places four 75-foot-long sections of precast concrete to form the roadway.

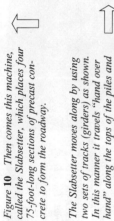

The Slabsetter moves along by using two sets of tracks (girders) as shown. In this manner it travels "hand over hand" along the tops of the piles and is on solid footing while it is performing its work, without ever "getting its feet wet."

Guardrail

Roadway

Figure 11 In the foreground is a man-made island that provides protection for the mile-long tunnel where it surfaces to connect with the trestle. The remaining three islands are visible in the background. Each island is 1600 feet long and required about two million tons of material. The island's sand core is protected by a heavy armor of stone.

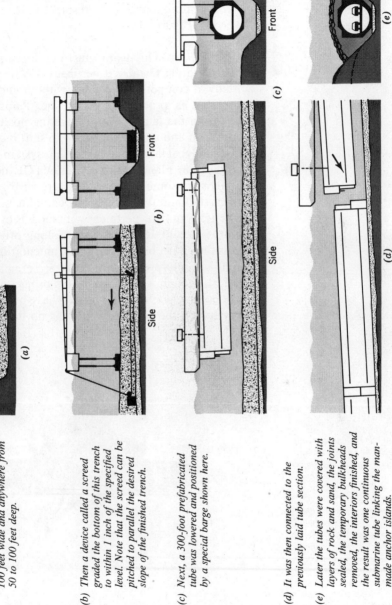

Figure 12 *The construction of the mile-long tunnels beneath the shipping channels is another interesting story.*

(a) *First a dredge cut a trench roughly 100 feet wide and anywhere from 50 to 100 feet deep.*

(b) *Then a device called a screed graded the bottom of this trench to within 1 inch of the specified level. Note that the screed can be pitched to parallel the desired slope of the finished trench.*

(c) *Next, a 300-foot prefabricated tube was lowered and positioned by a special barge shown here.*

(d) *It was then connected to the previously laid tube section.*

(e) *Later the tubes were covered with layers of rock and sand, the joints sealed, the temporary bulkheads removed, the interiors finished, and the result was one continuous submarine tube linking the man-made anchor islands.*

23

Aircraft Development For the last four years a seven-man design team has been developing a new type of airplane called a VTOL (pronounced veetol, meaning vertical take-off and landing plane). The results of their efforts are described in Figures 13-17.

Commentary. This team must develop a plane that has sufficient thrust to raise it vertically *and* to propel it horizontally at competitive speeds—without ending up with a plane that is practically all engine. Another challenge arises from the fact that in a hovering position this plane tends to tilt and cross winds tend to shift it out of position horizontally. Thus a sophisticated system is needed to maintain the plane's stability. In VTOL design stability problems are some of the stickiest, requiring much mathematical skill and frequent use of computers.

The team must experiment with models of various kinds (Figures 15 and 16). Therefore, among other things, this means they must be skilled in instrumentation, experimentation, and interpretation of data.

Oddly enough, even though this project has progressed to the point where working models have been test flown for many hours, the company still has no definite customer.

Figure **13** *A VTOL hovering in a stationary position. This plane can move straight up and down or hover like a helicopter by using the three lift fans—one in each wing and one in the nose. Once airborne, the pitch of louvers under the wing fans are changed to deflect the fan exhaust rearward, thus producing horizontal acceleration. When the aircraft speed is sufficient for normal aerodynamic wing-supported flight, the fan louvers are closed and the craft functions as a conventional, high-speed jet plane, cruising at about 500 mph. The fans are driven by the same two jet engines that propel the plane in horizontal flight. (VTOL photographs courtesy of Ryan Aeronautical Company.)*

Figure 14 A multiple-exposure photograph of the VTOL taking off, showing the transition from vertical to horizontal flight.

25

Figure 15 A ⅙ scale model of the VTOL being readied for wind tunnel tests. The tufts of wool attached to the fuselage show airflow. The cables leading into the rear of the model bring power in and information out from measuring instruments.

Figure 16 This is an experimental model of the type used in outdoor tests of preliminary VTOL designs. In this case the designers are evaluating the hover and control capabilities of a plane with three small jet engines at each wing tip. Eventually they removed the "leashes" and allowed this rig to rise several hundred feet. Crude as this flying framework may seem, it told the engineers what they needed to know—without unnecessary fanciness, time, and expense. (Courtesy of the Lockheed-California Company.)

Figure 17   The stakes are more than military contracts; naturally the company hopes to penetrate the commercial market with a VTOL. Therefore, the design team has been considering transport VTOLs like this one, based on the concepts they have developed for the military version shown in Figures 13-15.

27

The Army has financed the project, but the company has no guarantee that it will ever sell a VTOL. What the company receives in the future for this effort depends on how well these seven engineers do their job *and* on the designs competing companies come up with.

Ventures like this are becoming rather common. A company sees a distant opportunity in some type of engineering creation, which at present is considerably beyond the state of the art. It begins developing its technical competence in the area by assigning a team of engineers to design one or more experimental models, as was the case with the VTOL. Such development efforts are sometimes funded by a military or other agency; sometimes the cost is borne by the company with the hope that the investment will someday lead to profitable contracts. Usually a company carries on such projects in parallel with competitors, giving rise to the increasingly frequent "great engineering competitions" of the last few years. Examples: supersonic transport, jumbo jet, communications satellite, and computer traffic-control system for large cities. In each instance several companies are building technical competence that they gamble will eventually pay off. This adds an element of excitement as well as uncertainty to the project for the engineers involved.

Some Generalizations

An engineer is a problem solver. Typically his problem begins with the recognition of a need or want that apparently can be satisfied by some physical device, structure, or process. At this stage things are likely to be vague. For instance, the management of an automobile company has decided that it must be prepared to market an electric automobile—or be left in the competition's dust. The engineering staff has its assignment. In broad terms, management has specified the desired features of the new product, such as the approximate price range and power rating. The task remaining is to design a vehicle that satisfies the general performance characteristics given. This is typical of engineering assignments. The engineer is given the general function or purpose to be served and perhaps some loosely specified requirements and preferences for a solution. Such functional and performance specifications are usually selected by his superiors or client, often in collaboration with the engineer. Thus the engineer's prime task is to translate a loose statement-of-what-is-wanted into the specifications for a satisfactory means of fulfilling that objective.

Almost invariably there are numerous ways of accomplishing the specified objective, many if not most of which are unknown to the engineer at the start of the project. It is up to him to uncover and explore a number of possibilities. The knowledge that he has gained through education and experience is an important but not the only source of such solutions; he must also use his ingenuity. In evaluating the many possibilities he must rely heavily on his judgment, which he uses in lieu of an exhaustive investigation of all the alternatives (something he obviously can't take the time to do). This judgment, which comes with experience, is a demanding aspect of an engineer's day-to-day work. The creativeness required to invent solutions and the judgment used in evaluating them mean that the practice of engineering is more of an art than you may have assumed.

In almost every engineering project there is an air of urgency. Often a target date has been set for a solution, and usually there is pressure to produce results as soon as possible. As a consequence the engineer generally must recommend a solution long before he has had time to uncover all the possibilities.

The extent to which economic considerations enter into engineering work can hardly be overstated. If society is to benefit from the engineer's creations, they must be solutions intended users can afford. Moreover, private enterprise does not embark on a venture that does not hold high promise of yielding an attractive return on the investment. In public ventures too, a satisfactory benefit-to-cost ratio is demanded. Even though an engineer's proposal may satisfy its intended function admirably, that solution is doomed if it will not yield a net gain to the business or community. Therefore, the engineer must have a keen interest in costs—the cost of developing *and* of making *and* of using his solution.

An engineer must concern himself with the producibility of his creations, both from a technical (Can it be made at all?) and an economic (Can it be made at a tolerable cost?) point of view. The designers of the Chesapeake Bay bridge-tunnel carefully considered the effects of different routes and alternative structural features on the cost of constructing their creation. Similarly, the designer of the Diagnosticator is expected to specify a device that can be manufactured in quantity at a price the potential consumer will tolerate and at a profit to the business.

In most engineering problems there are conflicting objectives. The automobile manufacturer may want his electric auto to be comfortable, safe, powerful, light, and inexpensive, and to have a large carrying capacity—but he's not going to get it. The car cannot be the very best in *all* of these respects. If the designer does everything possible to maximize speed and power, sacrifices must be made elsewhere, probably in comfort, price, and capacity. And so it goes as he tries to make his design the ultimate with respect to any one performance characteristic. In the end the engineer must strike the best balance between conflicting criteria. This is no simple task.

Contacts with people require much more of the engineer's time, and sitting in solitude at the drawing board consumes much less, than is generally realized. A surprising proportion of an engineer's time is spent making inquiries, issuing instructions, answering questions, providing advice, exchanging ideas, and seeking approval. Consequently, inability to maintain satisfactory personal relationships can be a severe handicap to an engineer.

The engineer's involvement with humanity does not end here. An important part of his work is the detection and appraisal of human needs, for example, the need for new sources of fresh water, and the types, capacities, and number of water purifiers required. In addition, he must concern himself with public acceptance of his solutions and therefore must become familiar with the manner in which people will use his creations, the way in which they will react to them, and the features preferred by potential users. It is also his responsibility to anticipate and be concerned about the effects of his creations on the public at large, for instance, the impact of the bridge-tunnel on the lives of people in the community. Thus the engineer is deeply involved with social needs as well as with the acceptance and effects of his creations.

His involvement with people and with economic matters means that a greater proportion of his problems is nontechnical (but certainly no easier) than the layman realizes.

(Perhaps you should keep this in mind; it will help you understand why you must take a number of nontechnical courses.)

In general, the result of an engineer's efforts is something tangible—a physical device, structure, or process, as illustrated by the Diagnosticator, the desalinator, the reed switch machine, the bridge-tunnel, and the VTOL. This fact is probably the basis of a common misconception about engineering. Since the *result* is a device, structure, machine, or mechanism, people conclude that engineers spend the bulk of their time working with these things— like the mechanic or the television repairman or the laboratory technician. But this is not ordinarily the case. An engineer usually does much of his problem solving in the abstract. He works much more with information (e.g., fact gathering, computing, thinking, and communicating) than he does with things. Furthermore technicians are ordinarily employed to construct prototype models of the engineer's creations if these should be necessary, so that he has little occasion to "work with his hands." Therefore engineering work is quite different from what many people conclude that it is. And, more important, a young person who likes to tinker with automobiles, construct and repair electronic apparatus, or play with chemicals is probably no more likely to be successful or happy in engineering work than one lacking such inclinations.

Most of the creations described in this book are complex systems, in view of the thousands of components involved and the complicated interrelationships between them. As a consequence of this complexity, which is rather typical of engineering work today, and of the broad range of types of knowledge required in such projects, many engineering problems are handled by teams of engineers with varying backgrounds. The situation in which one engineer completely designs a device or structure is becoming rare (and such a person is seldom just out of college). In fact, hundreds of engineers are involved in the design of a spacecraft. They are divided into teams, one designing the propulsion subsystem, another the guidance subsystem, and so on for a dozen or so major subsystems.

As you see and read about remarkable and often exciting engineering achievements, do not conclude that all of the engineering work behind them is challenging and sophisticated. A certain amount of the activity is unglamorous, detailed, tedious dogwork. You have it in engineering and in every other occupation. Of course draftsmen and technicians will relieve you of some of it. Also, the computer is moving in rapidly to take over many of the repetitive chores engineers previously did "by hand." But escape them completely you will not.

(Suggested readings for this and other chapters are found on page 205.)

Exercises

1 *Take three familiar problems from home, school, or elsewhere, and define them in terms of states A and B.*

2 *What do you suppose an engineer does when he must solve a problem, yet there is no scientific theory on which to base his solution?*

3 *Take a familiar device or structure and attempt to identify some of the conflicting objectives with which the designer presumably had to cope as he arrived at his solution. (For example, in most cases he has to resolve this conflict: maximize the number and effectiveness of functions the device will perform but minimize the cost of manufacturing it.) For this purpose you might select a camera, a household appliance, or a power tool.*

4 *In almost every instance the engineer must propose a solution to a problem in a rather limited period of time. What do you imagine are the consequences of this restriction? (For example, what must he resort to, what sacrifices must be made, what must he forego that the perfectionist would like to do?)*

5 *Describe an engineering project, pursuing as far as possible the pattern set by the case studies in this chapter. You should include such things as the circumstances that gave rise to the project, the difficult or unusual problems encountered, the final result, and the ensuing benefits. Possible topics: relocation of the Temple of Abu Simbel; a high-speed, automatically controlled rolling mill for manufacturing sheet steel; San Francisco Bay Area Rapid Transit System; a computer-based airline reservation system; the jumbo jet; the Intelsat communication satellites; a nuclear power plant; and the Verrazzano-Narrows Bridge.*

3

THE ROOTS OF
MODERN ENGINEERING

MAN has always devoted much effort to the development
of devices and structures that make natural resources more
useful. He devised the plow to render the soil more pro-
ductive of food, the saw to transform the wood of the tree
to useful forms, the windmill to convert the forces of the
winds to useful work, the steam engine to transform latent
energy in fuels to mechanical work. These and a myriad of
other implements, machines, and structures are the results
of an unceasing search. In early times, as occupational
specialties were evolving, there arose, along with priests,
physicians, and teachers, experts devoted to creating these
devices and structures. These early engineers were responsible for creating weapons, fortresses, roads, bridges, ships,
and other contrivances. Their activity is readily traceable
to the ancient empires, and evidence of their remarkable
creations still remains, especially roads, aqueducts, and
defense works from Roman times.

These men were the predecessors of the modern-day
engineer. The most significant difference between these
classical engineers and their modern counterparts is the
knowledge on which their creations are based. The early
engineers designed bridges, machines, and other major
works on the basis of practical know-how, common sense,
experimentation, and inventiveness. The know-how was
an accumulation of experience passed on mainly by the
apprenticeship system, and to which each man contrib-
uted. In contrast to today's engineers these early practitioners had almost no knowledge of science, which is
understandable; there was practically no science to know.

Engineering remained essentially in this form for many
centuries. During the Renaissance the level of sophistica-

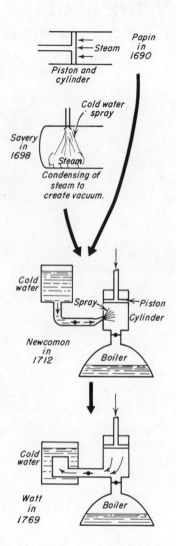

Papin in 1690
Piston and cylinder
Steam

Savery in 1698
Cold water spray
Steam
Condensing of steam to create vacuum.

Newcomon in 1712
Cold water
Spray
Piston
Cylinder
Boiler

Watt in 1769
Cold water
Boiler

tion increased, but even during the period of the steam engine's development in the eighteenth century the creators of machines and structures relied very little on science. Evolution of the steam engine illustrates the state of engineering during that period. The steam engine patented in 1769 by James Watt was one in a succession of progressively better machines, which began almost a century before. Watt made a significant improvement that vastly improved the efficiency of the steam engine and led finally to its widespread use. In the Newcomen engine, the predecessor to Watt's, steam that pushed the piston was condensed in the cylinder itself. This severely limited efficiency because on the up-stroke a hot cylinder was desired and on the down-stroke a cool cylinder was best. Under the circumstances it wasn't either. Watt added a separate condensing chamber and that change made all the difference. (Naturally, with the benefit of hindsight this improvement seems simple and obvious, but it was a long time coming in the eighteenth century.) Thus evolution of this machine is marked by a series of cumulative inventions by a number of men. Each relied on ingenuity, the contributions of his predecessors, and trial-and-error exploration sometimes over a period of years or decades. These engineers knew nothing about molecular activity, quantitative relationships between temperature and pressure, and other scientific facts.

Present-Day Engineering   The classical engineers were handicapped in what they could accomplish as long as they had little understanding of science, a situation that existed until relatively recent times. All this has changed. Within the last century and a half scientific knowledge has blossomed into an immense accumulation of information. Man's understanding of the structure of matter, electromagnetic phenomena, the elements and their relationships, the laws of motion, energy transfer processes, and many other aspects of the physical world has improved many times over. Much of what is now taught in high school physics was unknown when Watt developed his steam engine, and yet the contents of such a course constitute only a fraction of what is known today.

In the nineteenth century engineers recognized the potential that this growing body of scientific knowledge offers in the solution of man's practical problems and began to take advantage of it. With this major change—the vastly increased use of scientific principles in the solution of problems—classical engineering evolved into its modern form.

If you assume that contemporary engineering is simply an extension of science, as some writers mistakenly imply, you are missing a very important point and you do not have a true picture of the profession. Engineers existed long before there was any significant body of scientific knowledge, serving then, as now, as society's experts in the creation of its more complex devices, structures, and processes. Later, extensive improvement in man's understanding of the physical world brought about a significant change in this field. Present-day engineering is addressed to essentially the same types of problems, but science is now used extensively in the solution of these problems. Note, however, that inventiveness, expert judgment, and empirical knowledge are *still* relied on heavily in the solution of engineering problems.

There is a close parallel between the evolution of engineering and that of medicine. Specialists in the curing of bodily ills evolved in very early times. The predecessors of present-day physicians practiced through many centuries what was mostly art; there was no significant body of scientific knowledge on which to rely. In relatively recent times bacteriology, physiology, and other biological sciences developed into sizable bodies of knowledge, and physicians began to apply them in the treatment of health problems.

Thus both physicians and engineers are problem-solving specialists with roots deep in history, who eventually and logically have assumed the responsibility for *applying* a certain body of scientific knowledge. *They always were and still are problem-oriented.* Their prime motive is to solve a problem at hand. If perchance they are faced with a problem for which scientific knowledge does not supply a solution, they will still attempt to solve that problem. (The surgeon does not walk away from a patient on the operating table if he encounters a situation for which science does not tell him what to do!) The physician and the engineer have a job to do, and they will arrive at a solution to a problem through experimentation, common sense, ingenuity, or perhaps other means if current scientific knowledge does not cover the situation. Thus the engineer

*They do what they must;
use science when applicable,
intuition when useful, and trial
and error when necessary.*

　　　　　　　　Charles L. Best

does not exist *solely* for the application of science; rather he exists to solve problems, and in so doing he utilizes scientific knowledge when it is available.

The Distinction between Science and Engineering Full appreciation of the role of engineering is difficult if you do not understand the basic distinction between science and engineering. They differ with respect to *the basic processes characteristic of each (research versus design), predominant day-to-day concerns,* and *primary end product (knowledge versus physical contrivances).*

Science is a body of knowledge, specifically, man's accumulated understanding of nature. *Scientists* direct their efforts primarily to improving this understanding. They search for useful explanations, classifications, and means of predicting natural phenomena. In his search for new knowledge the scientist engages in a process called *research* and in so doing he devotes much of his time to the following activities:

- Hypothesizing explanations of natural phenomena.
- Obtaining data with which to test these theories.
- Conceiving, planning, instrumenting, and executing experiments.
- Analyzing observations and drawing conclusions.
- Attempting to describe natural phenomena in the language of mathematics.
- Attempting to generalize from what has been learned.
- Making known his findings through articles and papers.

Scientists explore what is and engineers create what has never been.

Theodore von Kármán

The scientist's prime objective is knowledge as an end in itself.

In contrast, the usual end product of the *engineer's* efforts is a physical device, structure, or process. Let there be no mistake—the gyroscope, the weather satellite, the radio telescope, the electrocardiograph, the nuclear power station, the electronic computer, and the artificial kidney are the fruits of *engineering.* The engineer develops these contrivances through a creative process referred to as *design* (in contrast to the scientist's central activity—research). Some of the engineer's prime concerns as he executes this process are the economic feasibility, the safeness, the public acceptance, and the manufacturability of his creations. In contrast, the scientist's prime concerns as he performs his function include the validity of his theories,

the reproducibility of his experiments, and the adequacy of his methods of observing natural phenomena.

Faraday's formulation of the principles of electromagnetic induction was a contribution to science. The use of this knowledge in the design of electric generators is engineering. When man came to understand nuclear fission in the late thirties this was an important scientific discovery. Applying this knowledge in the design of useful nuclear reactors is engineering. This is not to say that persons who basically are scientists never design instruments or solve problems, or that persons who we would call engineers do not do any research in the course of finding solutions to their problems. The key to the distinction is what constitutes a prime objective and what constitutes a means to an end. Engineers who developed practical means of converting salt and brackish water to usable form engaged in research to gain additional knowledge about the fundamental processes involved. However, they were engaged in this research in order to solve the problem at hand. Their goal was development of an economical conversion process.

Figure 1   This is by far the largest radio telescope. It is a wire mesh reflector supported in a natural hollow in the mountains of Puerto Rico. Radar signals originate from the movable transmitter and are reflected toward outer space. These signals bounce off planets and stars and return as echoes that are focused by the reflector on the receiver. These echoes are then analyzed by scientists to yield new knowledge about the universe.

Here is a prime example of an engineering creation the press typically calls a scientific achievement. To be precise, it is a scientific instrument used in radio astronomy, the design and construction of which are engineering achievements. Engineers determined the site for this telescope; they designed it; in fact it was an engineer who conceived the basic idea. I mention these things not because scientists and engineers quibble about credit due, but because the types of work involved in design and use of this instrument are quite different, which is important to young persons planning their careers. (Courtesy of the Cornell Center for Radiophysics and Space Research.)

When a space vehicle re-enters the earth's atmosphere at very high speeds, the heat generated will melt any known metal. It was necessary, therefore, for engineers designing re-entry vehicles to do some research to find a material capable of withstanding this intense heat. The knowledge that resulted is a by-product of their efforts to develop a successful re-entry vehicle.

Summary   Engineering as it exists today is primarily the outgrowth of two historical developments that until the mid-nineteenth century were essentially unrelated. One of these was the evolution ages ago of a specialist who has since served as society's expert in the creation of complex devices, structures, machines, and other contrivances. The other development is more recent: the accelerating growth of scientific knowledge. Although the marriage is a fairly recent one, it has already brought about a significant change in engineering. In contrast to the situation of the past, modern engineering involves more science and less art, although art in the form of creativeness and judgment is still conspicuous.

Exercises

1 *Prepare a paper on one of the following topics:*
   (a) *The contributions of engineering during the time of the Roman Empire.*
   (b) *The development of the steam engine through Watt's engine patented in 1769.*
   (c) *The bridges and aqueducts of Roman times.*
   (d) *The evolution of sources of power—slave labor to atomic energy.*
   (e) *The development of the dynamo, beginning with Faraday's discoveries.*
2 *Sooner or later more than one person will ask you to explain the difference between a scientist and an engineer. What will you say?*
3 *What is your reaction to this definition: "engineering is the application of science"?*

QUALITIES OF THE COMPETENT ENGINEER

YOUR success in engineering depends primarily on the *factual knowledge* you have acquired, the *skills* you have developed, your *attitude,* and your *capacity for continuing self-improvement.* This chapter describes what you should have in these four respects if you are to be a competent engineer. I hasten to emphasize that this applies to you in, say, ten years—not at graduation from college and certainly not now. I am describing what you should have to offer now *plus* what your college years add *plus* the benefits of some experience as a practicing engineer.

Now let me give you a good reason for becoming familiar with the qualities of a competent engineer summarized in Figure 1. An engineering education is designed to make a major contribution to your development in these areas. An understanding of this "big picture" is desirable because it enables you to be a more effective *partner* with your teachers and thereby to increase significantly the benefits of your education.

Factual Knowledge

A very important part of your formal education concerns the physical sciences, primarily physics and chemistry, as indicated by the number of courses in these subjects in the typical engineering curriculum. In order to develop complex devices, structures, and processes, an engineer must have a *fundamental* understanding of the laws of motion, the structure of matter, the behavior of fluids, the conversion of energy, and many other phenomena of the physical world.

Basic physical science

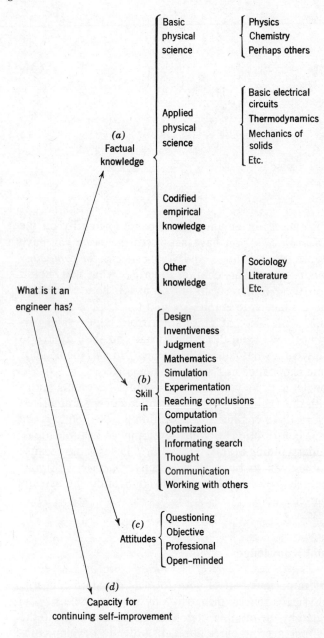

Figure **1**   *The "big picture" for this chapter.*

But knowledge of basic physical science is hardly enough. If an engineer is to solve problems, he must also study applied physical science and a body of codified empirical knowledge. I will elaborate.

Applied physical science

A body of knowledge pertaining to the "where" and

"how" of applying the principles of science is referred to as applied science. The beneficial application of fundamental scientific knowledge to the world's practical problems requires more than merely knowing the basic facts. When you are sick you don't want to be treated by a man whose only qualification is a knowledge of basic physiology and chemistry. Similarly for engineering; there is a big step between the basic principles of physical science and useful devices. The engineer's formal education must equip him to bridge this gap. Therefore, after you are familiar with basic physical science, you must take a number of courses devoted to the application of this science. An example of applied science is electrical circuit analysis, which concerns the application of knowledge of fundamental electrical phenomena (charges, electromagnetic waves, electron flow, etc.) to the understanding of basic electrical circuits. Other applied physical sciences are taught under such course titles as thermodynamics, mechanics of solids, fluid mechanics, and properties of materials.

This is mainly "communicated experience." It is difficult to conceive of an engineering creation that is based completely on scientific principles. Most designs are based partly on scientific knowledge and, of necessity, partly on experience and invention. Over a period of years, many ideas, practices, and observations, although not founded on scientific principles, have been shown through experience to be sound and generally useful. These have been recorded and perpetuated, and they constitute an accumulation of empirical knowledge that engineers rely on extensively. Part of your formal education is devoted to the study of this knowledge, ordinarily in design courses in the junior and senior years. Design courses are primarily concerned with the *application* of science and empirical knowledge in problem solving, as well as with the development of problem-solving skills and techniques.

Codified empirical knowledge

Specialization in Engineering   In practice it is customary to specialize to some degree, primarily because large and substantially different bodies of knowledge are required by different types of problems. It is virtually impossible for an engineer to be competent in designing bridges *and* television equipment *and* jet engines *and* metal refineries *and* textile machines. As a consequence, *some* specialization is inevitable. Thus during the latter part of your undergraduate program you will probably major in some branch of engineering. The choices are many; the major ones are described in Appendix A. It is primarily in the content of the upper-level design courses that education in the various branches of engineering differs. The student of electrical engineering studies the behavior and design of electrical machines, communication devices, power distribution systems, and the like, while the student of civil engineering learns about structures, water supply systems, city planning, and related subjects. Similarly students of the other engineering specialties concentrate on subject matter pertinent to their fields.

Although specialization along traditional lines is still common in engineering *education,* most problems encountered in *practice* require knowledge from two or more of the traditional engineering specialties, as demonstrated by the case studies in Chapter 2. Design of a chemical manufacturing process naturally requires considerable knowledge that is traditionally a part of the training of a chemical engineer, as well as some of the knowledge acquired by electrical, industrial, and mechanical engineers. As a result, an engineer must often work closely with other engineers who were educated in specialties different from his own, and must himself employ some of the knowledge from other branches of engineering. Thus the engineer typically finds that on the job his knowledge must extend across the traditional specialty boundaries. Mainly for this reason students are usually required to take some courses in engineering specialties other than their own.

Other knowledge   Note that there are a number of important nontechnical aspects of your intellectual development. To be professionally competent your knowledge must extend beyond physical science and engineering. It must include such areas as economics, government, psychology, sociology, and the humanities. This breadth of knowledge is important for a number of reasons.

- You must know the "economic facts of life." To be of value to your employer and of benefit to society you must be aware of the importance and intricacies of profits, costs, price-demand relationships, return on the investment, depreciation, interest charges on capital, and other economic realities. You will be constantly involved in economic decisions. To cope with these decisions effectively you must be as profit-and-cost conscious as the businessman himself.

- You must work with persons in many fields of endeavor, for example, economists, accountants, politicians, sociologists, psychologists, lawyers, and union leaders. You should be aware of the contributions these people can make; you should be able to talk with them intelligently, work with them, and understand their problems.

- A college education is preparation for more than making a living; it is preparation for living. Therefore your studies should not be concentrated entirely on science and engineering.

- Educational breadth equips and motivates you to show a real concern for the society you affect through your creations; there is no stronger argument for extending an engineer's education into the humanities and social sciences. The important matter of an engineer's social concern deserves treatment by itself, and so Chapter 14 is devoted to it.

Mainly for these reasons, at least twenty per cent of an engineer's undergraduate education is reserved for study of the humanities (literature, languages, philosophy, etc.) and the social sciences, such as sociology, history, and economics.

The Engineer's Skills

You will apply your knowledge with the aid of the skills, mainly mental, summarized in Figure 1*b*. Let me elaborate.

Skill in design

Suppose you are assigned to develop a new traffic control system for a city. You would do so through a process called design, the general procedure by which you convert the vague statement of what is wanted into the specifications of a system for fulfilling that purpose. Design is the very core of engineering; all that you bring to bear on a problem is done so through this procedure. Skill in the execution of this process is important enough to warrant five chapters in this book.

Inventiveness

Proficiency in design depends heavily on your inventiveness, so that this too is an important quality. Using it, you would conceive a number of traffic control schemes which

Good judgment

Mathematical ability

Simulation skill

Skill in experimentation

Measurement skill

Ability to reach intelligent conclusions

Computational ability

Skill in optimization

Ability to use information resources

you would subsequently evaluate in order to determine the best. This evaluation must usually be done while your ideas are still "on paper." You can see why; it would hardly be feasible to try out your alternative traffic control systems under actual conditions. Field tests would require too much money, time, and public patience. One method of predicting the performance of alternative solutions is judgment, another is mathematics, another is simulation (this is experimentation using a substitute for the real thing, like a wind tunnel test of a model airplane). You would probably use all three of these skills in the traffic control problem.

You must experiment, which means you should know how to prepare an experiment in order to obtain a maximum amount of reliable information with a minimum of time and expense. In experimentation and in many other phases of your work you will use your measurement skill. Closely related to measurement and experimentation is your ability to reach intelligent conclusions from observations. Even if measurements are of a simple nature, skillful interpretation of them is not as straightforward as you may think. This is so because of uncontrollable variation in the characteristics of all materials, objects, and devices, along with the facts that *no* measurement system is perfect and that most conclusions must be based on relatively small samples of observations. Such circumstances as these complicate the process of drawing conclusions. In general, the human being is notoriously inept at reaching conclusions, as he repeatedly demonstrates in the process of jumping to erroneous conclusions about his fellow men. The unsound conclusion-drawing tendency that you come by rather naturally is likely to carry over into your professional practice unless you train your mind to combat it. An important part of this is learning the many potential sources of error in drawing conclusions, the limitations of small samples, the roles of chance, uncertainty, and prejudice, and the importance of carefully evaluating the reliability of evidence available.

A digital computer is a mighty handy tool. The ability to use it, your slide rule, and other such aids constitute your computational skills.

You seek the optimum (best) solution. Optimization is a term applied to the process of arriving at the optimum solution; skill in this connection is important indeed.

As the mountains of available knowledge grow higher, the desirability and the difficulty of searching for information relative to a problem also increase. Your skill in the use of information resources is therefore becoming ever more important. You may think that there is nothing to this aspect of an engineer's work, but you can miss a lot of valuable information and waste a great deal of time if you are untrained in this respect.

Your thinking skills will not go to waste in any job that you may take. A major goal of an engineering education is to sharpen your reasoning, analytical, and other mental abilities. Although there are not many occasions on which these processes are openly discussed during an engineering education, a major objective of most courses is to contribute to the development of thought skills. The fact that these processes are seldom treated explicitly can be misleading; let there be no mistake; the "ability to think" is a highly salable commodity on the employment market.

Thought skills

Don't underestimate—as many embryo engineers do— the importance of your communication skills. You must be able to express yourself clearly and concisely if you aspire to be a good engineer. Probably the most effective way of convincing you of the importance of skill in oral and written expression, other than letting you learn by personal experience, would be to have you hear the many pleas voiced by employers as well as engineering graduates for *more* emphasis on these matters in college. Your communication skills include your ability to express yourself mathematically and graphically. Graphical skill, which is the ability to present information in the form of drawings, sketches, and graphs, is essential to successful expression of your ideas.

Communication skills

Your ability to work effectively with other people is obviously important. Engineering involves many contacts with many people; if you can't maintain cooperative working relationships with them you are in trouble.

Ability to work with people

There are other engineering skills, but those outlined are the major ones and should be enough to demonstrate that a number of skills are required for engineering practice. In Chapters 5 through 13 I will discuss five of these skills at greater length. Do not conclude that the remaining ones are unimportant. The skills chosen were selected for amplification because satisfactory introductory explanations of them have not been published, whereas there are useful books on experimentation, measurement, graphical communication, and other skills to which I have not devoted whole chapters.

The Engineer's Attitude

Certain qualities that you should bring to bear on problems are neither factual knowledge nor skills. Together they constitute what is best described as an attitude or point of view. They are summarized by Figure 1c.

Questioning attitude

Cultivate a questioning attitude, a curiosity for the "how" and the "why" of things. It will lead you to much useful information and many profitable ideas. Some of this questioning results from inquisitiveness, some from a certain skepticism that prompts you to challenge the profitability of some practice, the validity of some "fact," the advisability of a certain feature, or the necessity of a particular component. Questioning various "facts," requirements, features, etc., to make them "prove themselves"— especially when they are matters of long standing—can really pay off.

WHY?
WHY?
WHY?

Objectivity

In the course of a typical project you will be the focal point of biased opinions and special-interest pressures. Furthermore, you will be confronted by many situations that owe their existence to custom rather than reason. In the face of biases, pressures, and traditions, you must strive to be objective in making evaluations and decisions.

Professional attitude

You are expected to assume a professional attitude toward your work, the people you serve, those whom your solutions affect, and your colleagues, in the traditional manner of the professions. The professional person serves society as an expert with respect to some type of relatively complicated problem. Under these circumstances the layman trusts the professional person, and because of this confidence the latter has an obligation to perform his services ethically. Since most of your creations directly affect the well-being of many people, the public trusts that your designs will be safe and otherwise beneficial to the welfare of mankind. The public trusts also that it has received the full measure of service for which it has paid.

Your professional obligation includes more than living up to this trust placed in you by those whom you serve and affect. It also includes:

- Insistence on seeing a project through to successful implementation of the solution.
- A desire to follow up on that solution in order to benefit from experience with it.

- A willingness to keep abreast of the best practices and latest developments and to employ them.
- A feeling of responsibility toward your colleagues, expressed in your actions, your attempts to improve the status of your professional group, and your willingness to exchange "unclassified" information with those in your profession.
- Maintaining in strictest confidence the unpatented ideas, secret processes, unique know-how, etc., that provide your employer or client with competitive advantage.
- An urge to contribute to the betterment of mankind through your creations and advice.

No small matter in determining your value as an engineer is your open-mindedness to the new and the different. A flexible mind is a real asset. Be receptive to new theories, new ideas, and innovations in technique.

Open-mindedness

The Capacity for Continuing Self-Improvement

When you emerge from college, you will not have all the characteristics described in this chapter. *The receipt of a baccalaureate degree in engineering marks the end of the beginning.* Your formal education provides a sound start in a long-term development process. After that it is up to you to continue your intellectual development if you aspire to blossom into a real engineer and to enjoy a fascinating, rewarding career.

As knowledge continues to accumulate at an increasing rate and technical problems become more complex, the ability and the inclination to continually build on what you have learned in college become more important. The means for this continued growth are experience, books and journals, conferences and workshops, trade publications, and postgraduate courses. Even if you were a finished product at graduation you would still find it necessary to continue your learning efforts because many of the things learned in college become obsolete within a relatively few years. By simply "standing still" you will eventually be little more than a "handbook engineer" (so called because the relatively routine problems given to him can be solved mainly by thumbing through handbooks and catalogs). The need to keep yourself up to date is growing because the rate of turnover of knowledge and techniques is steadily increasing.

Of course, continued self-improvement, in addition to being a professional obligation, makes good financial sense. Salary and rate of advancement depend partly on your inclination and ability to keep in tune with the times.

Objectives of an Engineering Education

Looking at Figure 1, you can anticipate the major objectives of an engineering education:

1. To impart a significant part of the factual knowledge you will require.
2. To give you a substantial start in the development of your engineering skills.
3. To help shape your attitudes.
4. To equip and motivate you for continuing self-improvement.

Which courses and parts of courses contribute to the first objective will be obvious. This is true also for the development of some skills, such as those learned in slide rule instruction or graphics courses. But attempts to develop your thought skills, ability to work with others, objectivity, open-mindedness, and other attitudes are not so obvious. Although these matters are seldom treated explicitly, your development in these subtler respects is an objective of most of your teachers.

The Value of an Engineering Education

The qualities summarized in Figure 1 are chiefly responsible for the reputation engineers have achieved as problem solvers. Your success in acquiring these characteristics determines your effectiveness as an engineer and the rewards you receive from a career in this profession.

An engineer's familiarity with science may be concentrated in the physical sciences but certainly is not limited to them. This familiarity extends to the social sciences and may include the biological sciences as well. Thus the formal education of an engineer typically spans the natural and social sciences, technology, and the humanities, making it a broad education indeed. This is significant, for breadth of knowledge *is* desirable, and today an education cannot justifiably be called broad if it does not include technology. In our civilization technology has become a potent force deeply affecting business, government, education, the military establishment—you name it.

Note that skills and attitudes cannot be acquired in the same manner as facts. Does a person suddenly develop a questioning attitude upon reading what such an attitude consists of and that he should possess it? It takes more than this—a disciplining of the mind over an extended period is required. Note too that, although skills and attitudes require more time and effort to acquire than factual knowledge, they are just as difficult to lose. Furthermore, specialized scientific and technical knowledge is susceptible to obsolescence as new discoveries are made. And too, if you change occupation, you may not need the specialized factual knowledge that you have gained, but the skills and attitudes acquired from an engineering education will be valuable in almost any field. These are "general-purpose" benefits.

Summary Description of Engineering

Chapters 2 through 4 provide the background for this definition: engineering is the application of certain knowledge, skills, and attitudes primarily in the creation of physical contrivances that serve society's needs and wants.

Engineering's emphasis on the *application* as opposed to the generation of knowledge was stressed in Chapter 3. The *knowledge, skills, and attitudes* of engineering are described in this chapter. Engineers are creators primarily of *physical contrivances*—tangible devices, structures, and processes. They are responsible for the *creation* of these things, that is, designing them and overseeing their construction. These contrivances are produced in response to society's *needs and wants* (it is naive to think we cater only to needs).

Exercises

1 *Assuming that you are in engineering school, each course that you are now studying is intended to contribute to your development with respect to certain of the qualities de-*

scribed in this chapter. Analyze the content and conduct of each of these courses; isolate the types of knowledge, skills, and attitudes that the course is apparently intended to develop in you.

2   *Prepare a list of devices, structures, and processes created by each of the major branches of engineering described in Appendix A.*

3   *Prepare a list of fifteen engineering creations the design of which probably required the talents of engineers from two or more major branches of engineering. Identify the branches that you believe were involved in the development of each creation that you name.*

4   *Sooner or later your girl friend is going to ask innocently, "Just what is engineering?" Since you surely will not have this book at hand to read from, your explanation will of necessity be expressed in your own words and will be (no reflection on your girl's intelligence) "down to earth." How will you explain engineering to her?*

5

MODELING

Iconic, Graphic, and Diagrammatic Representations The following objects have something in common: toy train, global replica of the earth, statue, model of an airplane. Each is a three-dimensional *representation* of a physical reality. There is a two-dimensional equivalent, as exemplified by the photograph, sketch, or blueprint. Since these two- and three-dimensional representations bear a physical resemblance to their real-life counterparts, they are referred to as *iconic representations*. Engineers make frequent use of iconic representations; you can see why the one pictured in Figure 1 is a valuable aid to persons who are trying to visualize what this structure looks like from 27 pages of complicated blueprints.

Then there is the familiar *graphic representation,* such as those on pages 61 and 62. You are already familiar with the usefulness of graphs in aiding you to visualize relationships and relative magnitudes.

A diagram usually represents some real-life counterpart. The schematic diagram for an electrical circuit and Figure 2 are *diagrammatic representations*. Other examples appear on pages 60 and 71. In each instance a configuration of lines and symbols represents the structure or behavior of a real-life counterpart. A diagram like that in Figure 3 is certainly helpful to the designers of the system it represents, especially since there are so many parts and interconnections. As you can imagine, with engineering devices, structures, and processes becoming so much more complex, diagrams must be relied on extensively in designing these systems and in communicating their make-up and operation to others.

Mathematical Representations The mathematical expression in the margin is a representation. The letter m represents the mass of gas present, T represents its tem-

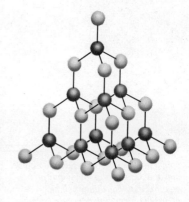

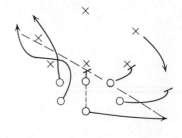

$$V = \frac{mkT}{p}$$

Figure 1   An iconic representation of the proposed design for an oxygen-production facility. This table-top model aids the designers in laying out piping especially so that things will be accessible for maintenance and repair. It is also used as a general visualization aid to them and to persons to whom they must explain their design. (Courtesy of Air Products and Chemicals, Inc.)

perature, p represents the pressure being applied, and V represents the volume occupied by the gas. Together these letters represent what happens to one of these properties when a specified change occurs in another. This *mathematical representation* provides a means of predicting one property, given specific values of the other three, for example, of predicting V for particular values of m, T, and p. Through the power of mathematics predictions can be made of many other natural phenomena as well as of the behavior of man-made mechanisms, structures, and processes. By employing the system of rules and conventions prescribed by mathematics, and by assigning symbols to represent relevant properties of the real thing, mathematical expressions can be manipulated to make useful predictions of what is to be expected under given conditions.

Mathematics provides a repertory of ready-made mathematical representations (parabolic function, exponential function, etc.). Training in mathematics also equips you to derive special expressions which you tailor to fit situations that cannot be satisfactorily represented by ready-made mathematical functions. This is a very important skill.

Mathematics is a powerful method of representation. It is an effective means for *prediction* and a concise, universally understood language for *communication*. Its conventions make it an extremely useful *medium for reasoning*. Can you imagine trying to perform in words some of the logic and the manipulations that you conveniently carry out through the symbolism of mathematics? In addition, training in mathematics has a beneficial effect on your ability to think clearly and logically. In view of the great utility of mathematics as a means of prediction, communication, and reasoning, the heavy emphasis given to this subject in engineering education is understandable.

Simulation An iconic representation can be used to predict the behavior of its real counterpart. A model of a proposed aircraft is subjected to high-velocity winds in a wind

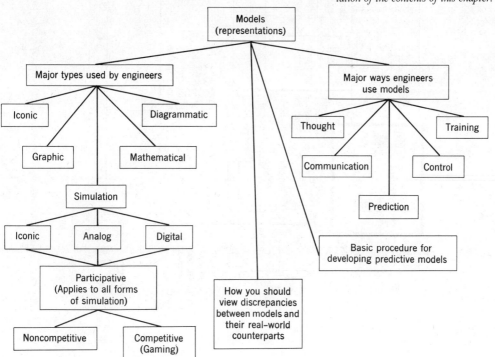

Figure 2 A diagrammatic representation of the contents of this chapter.

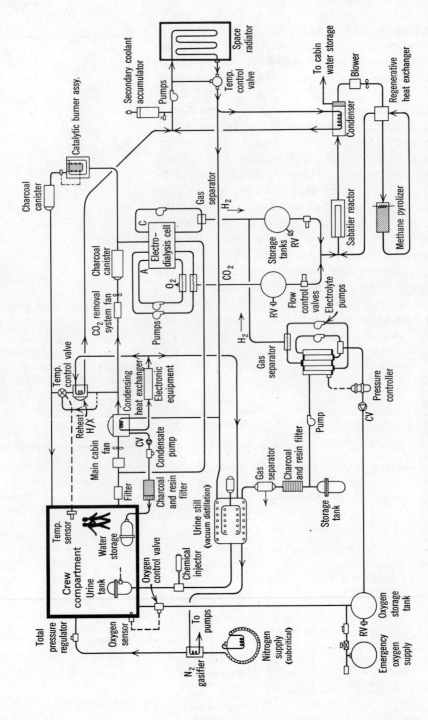

Figure 3 A diagrammatic representation of a life-support system for prolonged space missions. (Courtesy of Lockheed Missiles and Space Company.)

tunnel in order to predict how a real plane of that design will perform in actual flight (page 26). What the wind tunnel and plane model are to the aircraft designer, the facilities pictured in Figure 4 are to the designers of ocean-going vessels, and the facilities shown in Figure 5 are to

Towing carriage   Technician   Ship model

Figure 4   This is a basin in which models of ocean-going vessels are tested for maneuverability and seaworthiness. The carriage that tows and maneuvers the models rides along a 376-foot bridge which itself is movable. Special machines generate waves of specified size and frequency. This simulation setup enables engineers to predict the full-scale performance of proposed ship designs at sea. (Official photograph, U.S. Navy.)

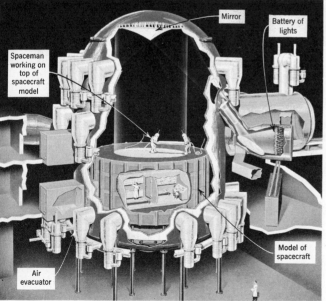

Mirror   Battery of lights

Spaceman working on top of spacecraft model

Model of spacecraft

Air evacuator

Figure 5   A cutaway view of a large chamber in which some conditions encountered in outer space are artificially recreated. In this scene a life-size, prototype model of a large manned space vehicle is under test in the chamber. (One outer-space condition that is ignored here is weightlessness, which as you can imagine is not readily recreated.) (Courtesy of the Lummus Company.)

Figure 6   This is a part of a working scale model—the largest in existence, covering over 200 acres—representing the Mississippi River and its tributaries from Sioux City, Iowa, to the Gulf of Mexico. This simulator is used to predict the local and system-wide effects of proposed dams, diversionary channels, and other construction projects. (U.S. Army Photograph.)

the designers of spacecraft. This process of *experimenting* with a *representation* of the real thing is called *simulation*. When the experiments are performed on iconic representations like those in Figures 4–6, the process is called *iconic simulation*.

There are two other forms of simulation, but in these instances the representations on which the experiments are performed bear a behaviorial but not a physical resemblance to their real-life counterparts. One of these is called *analog simulation*, the other *digital simulation*.

An example of analog simulation is the electronic device used by an engineer who is designing a traffic control system. Special electrical circuits represent the city's traffic arteries, while electrical pulses represent vehicles. With this simulator the engineer experiments with different traffic control schemes. Here electrical pulses behave analogously to autos moving about the city, even though pulses and wires in no way physically resemble autos and streets.

In the analog simulator pictured in Figure 7 water behaves analogously to air. It enables designers of gas turbines to test their ideas quickly and cheaply. Thus in analog simulation a medium that behaves analogously to the real phenomenon is employed as a vehicle for experi-

Figure 7   An analog simulator. In this instance water behaves analogously to air and serves as the medium for experimentation. The device simulates air flow through the diffuser blades (represented by wedge-shaped objects) in the compressor stage of a gas turbine under development. Water containing a dye which makes the path of flow readily observable is diffused from the center at one thousandth of the velocity of the gas it represents. By experimenting with different wedge shapes, angles, and locations, the investigators will learn how to maximize the effectiveness of this part of the engine. (Courtesy of the Boeing Company.)

mentation. Electricity is the medium often used. For example, voltage might represent the pressure of steam in an electrical analog simulator of a steam power plant.

Digital simulation is best introduced by an example. A university plagued by auto parking problems has engaged a consulting engineer to improve the situation. One of his ideas is to separate the drivers who consistently use their parking spaces the full working day (henceforth referred to as the "9 to 5 users") from those who use their spaces sporadically (e.g., a few hours one morning, all the next afternoon, an hour the following morning, etc.). The 9 to 5 users will be assigned regular spaces, but the sporadic users will be grouped and assigned to a lot where they will park in any available space. The engineer theorizes that more of the sporadic drivers can be assigned to a lot than there are spaces, with negligible risk of the lot overflowing. Notice, though, that this is an untested idea. How does he go about verifying his hypothesis?

One way is to set up a special lot, assign only sporadic users to it, and then observe it for a significant period of time to see what happens. This is too costly and too involved and will take too long. A logical alternative to this real-life experiment is to test the theory by simulation, which the engineer did. In this simulation he assumed a 20-space lot to which 25 drivers were assigned. His procedure and results are described in Figures 8–10, which you should consult before reading on.

TABULATION SHEET FOR PARKING SIMULATION

Day No. ___1___

Identification of Row	Source of Table Entries	1	2	3	4	Driver No. 5	6	7	21	22	23	24	25
Show or no show?	Spinner	Yes	Yes	Yes	No								
T_a (arrival time)	Spinner	8.5*	10.0	8.9									
P_e (expected park time)	$P_e = 10.2 - .56\,T_a$ (or use graph)	5.4	4.6										
D (chance deviation)	Chip drawn from bowl	+.5	−1.1										
P (actual park time)	$P = P_e + D$	5.9	3.5										
T_d (departure time)	$T_d = T_a + P$	14.4 (2:24)	13.5 (1:30)										

* To perform these numerical operations the hours must be in decimal form: 12 midnight = 0, 9:30 A.M. = 9.5, 3:00 P.M. = 15.0.

Figure 8 Here is the crux of the digital simulation. In executing this simulation the operator performs the indicated numerical operations, using this sheet as a running record of the experience he has synthesized, until he has generated a significant number of data.

This graph is a summary of 100 daily maximums. Each daily maximum is determined as explained in Figure 10. Here's where the value of the simulation becomes apparent. Note that on only one of the 100 days simulated did the lot overflow, even though 25 drivers were assigned to a 20-space lot. Furthermore, over this period, on only four percent of the days did the daily maximums exceed 19 autos, and on only 8% of the days would the lot have overflowed if there had been 18 spaces. From this it is apparent that more drivers of the sporadic type can be assigned than there are spaces on the lot, resulting in better utilization with negligible inconvenience to the users. Furthermore, the client can decide what probability of overflow he will tolerate, then the engineer can use this graph to determine how many drivers can be assigned to a given lot without exceeding this probability. Suppose university officials say that they don't mind if the lot overflows 25% of the time. For this not to be exceeded, 16 spaces should be provided for a group of 25 drivers of the sporadic type. (Don't let this 25% alarm you; even when the lot overflows only one or two drivers are usually affected and there are other lots for them to go to.)

Thus by simulation the engineer was able to evaluate the effects of a policy of overassignment before making any *real* changes, *and* he had something to back up his proposal when he offered it to his client. It took a clerk three days to perform this simulation; gathering data and setting up the procedure required another two. Thus in five days it was possible to synthesize 100 days of experience, illustrating the "time compression" capability of simulation.

Digital simulation is experimentation with a digital model. It is an "acting-out process" in terms of numbers that is simple yet remarkably powerful. Because it is a series of step-by-step numerical operations, it can be executed by a computer, as explained in Appendix B. This is fortunate since performance by pencil and paper is laborious and time consuming. In fact, without this machine to do the highly repetitive numerical manipulations digital simulation would be prohibitively expensive in many potential applications.

Digital simulation by a computer has recently become very popular in engineering practice. Its applicability is illustrated by this sample list of uses: simulation of the behavior of atomic particles, of space flights, of air and ground traffic, of economic systems, of global warfare.

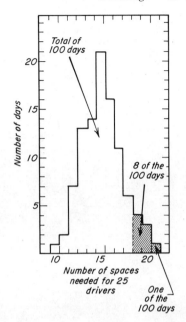

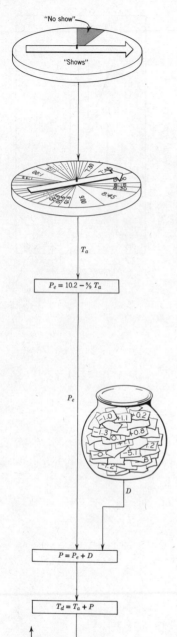

From a survey the engineer estimates the probability that a driver will not use the lot on any given day as 8 chances in 100. This "no show" probability of 0.08 is taken into account by using a roulette-type device which has a free-spinning pointer (left). Beneath the pointer is a pie chart (right) having a "no show" sector of 28.8 degrees (0.08 × 360 degrees). This pointer is spun for each of the 25 drivers, to determine whether he uses the lot on that day. If the pointer stops above the "no show" sector, this fact is recorded on the tabulation sheet and the device spun again. If it stops above the "yes" sector, on to the next step.

The likelihood of a driver arriving is not the same throughout the day. To determine how arrival frequency changes with time of day some actual observations were made. These data (a, at right) were converted to a pie chart (left) in which each sector is proportional to the equivalent bar in the graph at right. In the simulation this pointer is spun to select an arrival time (T_a) for each driver, and this value is recorded on the tabulation sheet.

The length of time (P) an auto remains on the lot depends on the time it arrives. To learn the nature of this relationship some "park times" were measured (b, at right). A straight line was fitted to this data. From the equation of this line the *expected* (most likely) park time (P_e) can be computed, given the driver's time of arrival. (P_e is the average parking time for many drivers arriving at a given time of day.) P_e is calculated for each driver after his time of arrival is generated.

From Graph (b) it is apparent that the *actual* periods of time autos remain on the lot vary considerably from the *expected* times (i.e., from the straight line). This is taken into account in the simulation. The vertical deviation of each point from the straight line (graph b) was measured (plus if above the line, minus if below it). Graph c at right summarizes these measurements. Each of these deviation values was recorded on a separate slip of paper and placed in a bowl. In the simulation a slip is selected randomly, the D value noted, and the slip returned to the bowl.

Then this D value is added to the *expected* park time (P_e) to obtain the *actual* park time (P).

The departure time (T_d) of a given driver is calculated by adding his park time to his time of arrival.

Figure 9   *This sequence of steps is repeated 25 times to complete the simulation of one day of parking-lot operation. Then another day is simulated, and another, until a sufficient amount of experience in the operation of this hypothetical parking lot has been synthesized. The results for one simulated day are shown in Figure 10.*

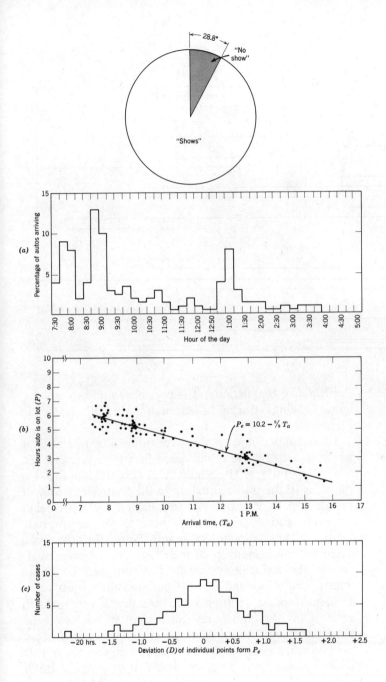

The spinner device is used to recreate the element of chance associated with events. The use of this device and the drawing-from-the-bowl procedure are equivalent. These and other ways of making random selections from a collection of numbers are appropriately referred to as Monte Carlo Procedures.

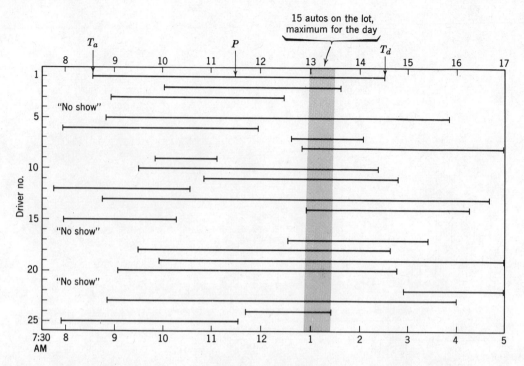

Figure **10** *A graphic representation of one day of simulated parking-lot operation. The maximum number of autos is determined by noting the maximum number of simultaneously overlapping bars.*

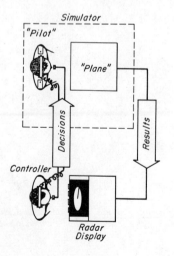

Participative Simulation It is possible to *directly* involve human beings during a simulation, as illustrated in Figure 11.

These "pilots" and "planes" constitute a large simulator which is used by the traffic controllers to simulate traffic control operations. The whole thing operates like this. On the basis of the traffic situation in the controller's zone of jurisdiction, as it appears on his radar screen, he makes decisions and communicates them to the appropriate "pilots." They respond by manipulating their controls. The resulting maneuvers of their "planes" become apparent on the controller's screen. The controller continues with his decisions and instructions, and the "pilots" continue to respond, creating a situation that is very realistic, at least for the controllers. Such simulations have been used, for example, to predict the impact of a *proposed* jetport on existing air traffic lanes and to predict the impact of supersonic transports on traffic flow and on traffic controllers.

Many simulations directly involve people as decision makers in this same general manner, as exemplified by Figures 12 and 13. (Another example appears on page 195.)

Figure **11a** *An air-traffic control center. In front of each controller is a radar screen on which the usual pips appear, representing the positions, identities, and altitudes of the planes in his area of control. Each controller is radioing instructions to planes in his zone of responsibility and receiving reports from them. To the casual observer this scene looks real enough. However, the "pilots" with whom these controllers are communicating are in an adjacent room—Figure 11b.*

Figure **11b** *Pilots and planes? Well not really, but as far as the traffic controllers shown in Figure 11a are concerned, these are. Each person in this room is playing the part of a pilot, and the special apparatus in front of her is a "plane". As she "flies" her flight plan (e.g., destination, course, altitude) by manipulating the controls on the console, this information is sent by wire to a controller's radar set, where it causes a pip to move realistically across the screen. The usual controller-pilot conversations take place over telephone lines between the two rooms. (Courtesy of Federal Aviation Agency.)*

Figure **12a**   *This astronaut, shown maneuvering his spacecraft during rendezvous with another vehicle, can go home at 5 PM with other employees of this aerospace company because he is performing this mission on a simulator in the laboratory.*

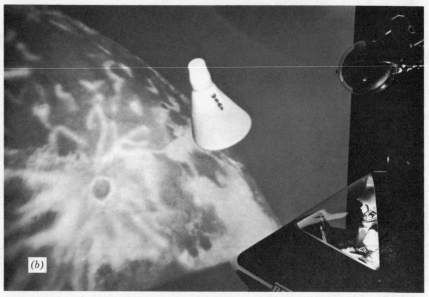

Figure **12b**   *The effect pictured in (a) is created by separately projecting moon and spacecraft images onto a large semi-circular screen in front of the vehicle from which the scene in (a) was photographed. As the astronaut manipulates his controls, the scene in front of him changes appropriately, providing him with the illusion of movement with respect to the moon and to the other vehicle. In an adjacent room a TV camera, focused on a 3-dimensional model of the moon, picks up the image that appears on this large screen. The camera moves in response to the astronaut's manipulation of the controls. A similar closed-circuit TV system projects the vehicle on the screen. Launches, landings, and orbital missions can be simulated in order to learn what astronauts can and cannot do. (Courtesy of the Boeing Company.)*

(a)

Cockpit

Lights

(b)

Figure **13a** *An airport runway at night? Well, not exactly, but it certainly looks real.*

Figure **13b** *The scene in (a) was taken in this building at night from the "cockpit" shown. The cab moves up and down, back and forth, and lengthwise in response to the "pilot's" manipulations of the controls. The runway lights visible in (a) are located on the floor of this room. This 1000 foot-long facility is used to simulate landings in fog, in order to predict the effectiveness of lighting and other guidance systems. A spray system creates a fog of desired density. (Courtesy of University of California Institute of Transportation and Traffic Engineering and the Federal Aviation Agency.)*

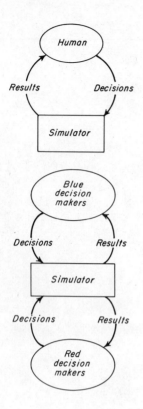

Other illustrations: in practice warfare commanders communicate their decisions to a computer and receive from it the outcomes of those decisions; pilots, missile launch crews, drivers, etc., practice their tasks on simulators. The pattern is the same in each instance: the human being makes decisions and communicates them to the simulator, which returns the results of these decisions via visual displays, indicators, etc. On the basis of these results the person makes new decisions, and this cycle of events is repeated over and over.

Since a human being participates directly, I call this *participative simulation.* If two or more persons compete with one another in a participative simulation, it is referred to as *gaming.* Participative simulation is useful for prediction and for training purposes. Most of us would prefer to have a fledgling pilot make his learning mistakes on a simulator that realistically reproduces actual flying controls and conditions. The consequence of a miscalculation in actual flight is likely to be infinitely worse than the mere sounding of a loud buzzer.

Thus an engineer can experiment on iconic, analog, or digital representations in order to make predictions about their real-life counterparts. This process, simulation, is a means of synthesizing experience by operating a model for a period of time to learn how the real thing would perform. This usually beats experimenting with the real thing in several respects; it costs less, takes less time, and enables engineers to exert closer control over their experiments.

Models The representations described in this chapter are usually referred to in the literature and conversation of engineers as *models.* The engineer ordinarily speaks of graphic models (instead of graphic representations), diagrammatic models, etc. This requires that you extend your interpretation of the word model beyond the ordinary concept. To the engineer a *model* is anything that describes the nature or behavior of a real-life counterpart. It can do this describing through words, numbers, symbols, diagrams, graphs, or by looking or behaving like whatever it represents.

It may take a while for you to fully appreciate the generality of this concept. When it does dawn on you, it will be clear that all of the following are models in the sense this term is used by scientists and engineers: a mental image of a person or of an experience or of anything; man's conception of the nature of light; Darwin's theory, Einstein's theory, or *any* theory; the words dog, stick, book, stone, or almost *any* noun; a verbal description of the workings of a mechanism; a musical score; a chemical formula. Then you will not be puzzled if you hear someone refer to Darwin's model, the wave model of light, or your model of engineering.

How Engineers Use Models

For Thinking A model can be a real help when you are trying to visualize the nature or behavior of a system or phenomenon which the unaided mind finds difficult to grasp. There are electrical circuits, manufacturing systems, chemical processes, and mechanisms so complex that a diagrammatic or other type of model is essential to human comprehension. Iconic, diagrammatic, and graphic models are especially helpful in providing a compact, overall, simplified view of the whole. Often, in thinking of a physical phenomenon, the engineer finds it expedient and profitable to consider it in terms of a model. For instance, an experienced engineer usually thinks of an alternating electrical current as a sine wave in graphic form rather than as the movement of electrons in a conductor. You might think of a gust of wind in the vague terms of a chilling sensation on your face; but an engineer who designs aircraft or long-span bridges will probably think of it in graphic form. Often it is these abstractions that engineers manipulate in their thinking. An objective of an engineering education is to develop your ability to think of physical phenomena in terms of useful abstractions.

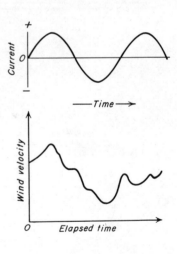

For Communication Of course all communication is achieved via models. I am communicating to you through symbols, photographs, sketches, and diagrams. In addition to these common communication aids, engineers frequently use mathematics, graphs, and working models, especially when the systems and phenomena they wish to communicate are complex. Imagine trying to describe in words the system diagrammed on page 54.

$$T_\mathrm{w} = \frac{T_\mathrm{u}^2}{T_\mathrm{a} - T_\mathrm{u}}$$

where

T_w = *average length of wait*

T_a = *average time between barge arrivals*

T_u = *average time to unload a barge*

$$T_\mathrm{w} = \frac{(3.1)^2}{5.7 - 3.1} = 3.7 \ hours$$

For Prediction   The designer of a rail-air-marine freight terminal must predict how long incoming barges will have to wait for unloading in order to determine whether one dock will be sufficient. This would be a trivial problem if the time of arrival and the unloading time of each barge were known. But they are not; arrival times are unpredictable, and unloading times vary considerably from barge to barge. Hence the engineer is relying on a mathematical model (margin) for his predictions. From his data he estimates that the *average* interval between barge arrivals will be 5.7 hours and the *average* unloading time will be 3.1 hours. Substituting these values into his model, he computes the average waiting time to be 3.7 hours. The engineer is using this model to predict the consequence, in terms of barge delays, of providing one unloading dock. He can do similarly for alternatives, such as providing a second dock or adding equipment to reduce unloading time, in order to find the best solution. Of course, all this is done while the terminal exists only on paper, and herein lie the beauty and the power of the predictive model.

The case just described is typical; in solving problems engineers must evaluate most alternative solutions while they are still in the conceptual stage. Models are extremely useful for this purpose; they enable the engineer to make the required predictions of solution performance without the necessity of physically creating the solution. Through the manipulation of mathematical and simulation models it is possible to evaluate solutions with less time, cost, and risk than experimentation with the real thing ordinarily requires, yet with greater accuracy than is usually possible if pure judgment is used. It is hardly feasible for the engineer designing the freight terminal to construct a full-scale version of each system he is considering so that he can experiment in an effort to determine which is best. And can you imagine how rapidly the aerospace industry would consume test pilots if it did not utilize simulation so extensively. In this and many other situations the costs of experiments on the real thing are very costly, yet the stakes are too high to rely solely on opinion. Predictive models are an excellent compromise in such situations.

For Control   When developing a model for prediction purposes the engineer wants the model's predictions to agree as closely as feasible with what eventually occurs. In some situations, however, the converse is true; a model is developed and an attempt is made to force the represented situation to conform to it. The engineering drawings for a building constitute a model, and of course the building is constructed to conform to that model. The flight path a spacecraft must follow to reach its objective is carefully computed beforehand. This planned flight path is a model, and very elaborate systems are employed to hold the actual flight path to this model.

For Training Most models that are useful for communication are useful also for instruction. Not so obvious though are the value and growing popularity of participative simulation as a training tool, especially where the investment in equipment and/or the probable consequences of blunders are very high. This is the reason for using simulation so extensively in training pilots, air traffic controllers, and astronauts. The astronauts and all key ground personnel repeat their space feat many times through simulation in the laboratory before the real countdown and blastoff. Similarly, the crews that operate our missile defense and attack systems get negligible opportunity to practice the real thing, yet they must become proficient at their tasks. The answer is simulation.

In summary, the concept of the model is powerful both in its utility and in its unifying ability. It logically relates many of the important subject areas that constitute an engineering education and should assist you to visualize the importance of various courses to the practice of your profession. The engineer's ability to capitalize on this technique for thinking, communication, prediction, training, and control is important indeed.

Model Versus Real World

Some realities associated with the operation of a campus parking lot are ignored—"assumed away," as the saying goes—by the simulation model previously described. As far as the model is concerned, weather has no effect on the probability of a driver using his car, yet we know this is not true in real life. Furthermore, some features of the parking lot model are not true of the real thing. For example, it was possible in the model to generate a negative stay time (it

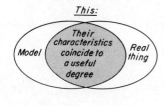

This:

Not this

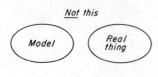

Or this

was ignored if it occurred), which obviously does not occur in real life. Such _discrepancies_ between a model and its real-world counterpart are inevitable. They are found in _every_ model.

Similarly there are such discrepancies in the case of the model used on page 68 to predict the average delay experienced by barges waiting to be unloaded. According to the model, the unloading crew does not work faster when the number of waiting barges becomes large, and the probability of a barge arriving is the same at all times of the day. We know these conditions are not true in real life. In the case of the equation given on page 51, called the equation of state of an ideal gas, there are differences between the ideal gas and a real one. _Yet these models provide useful predictions._ The fact that discrepancies exist between model and real thing, by itself, is unimportant. What matters is whether or not the final result—the prediction—is satisfactory for the particular purpose at hand.

Simplifying assumptions are made for good reasons. In many cases, if some of the complicating factors are not assumed away, it is virtually impossible to develop a usable predictive model. Furthermore, many discrepancies are of negligible _practical_ consequence; to remove them would make the model more complicated and costly and do virtually nothing to improve accuracy. Some of the discrepancies between the parking lot model and the real thing could be removed, but the model would then take longer to set up and run and therefore be more costly; the small increase in accuracy would not justify these disadvantages. Also, we could add seats and other fixtures to an aircraft model prepared for wind tunnel tests, but these have no effect on the aerodynamic characteristics of the model; a solid replica suffices and is certainly cheaper. Complicating, costly factors that are irrelevant or inconsequential are ignored in model construction and understandably so.

Developing Predictive Models The following general procedure for the development of a predictive model is of great significance in science and engineering. It is summarized in Figure 14.

1. A model that is _potentially_ satisfactory for the prediction task at hand is developed or selected. It can be a model prepared especially for the problem at hand, such as the

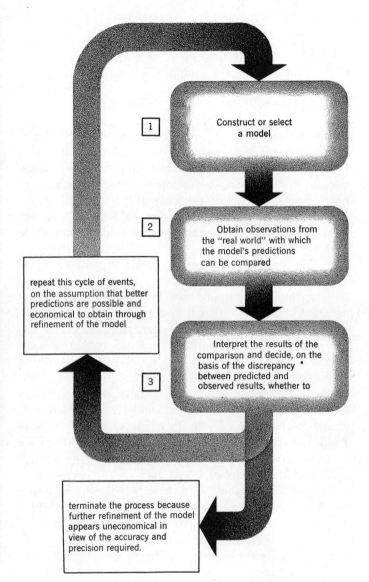

Figure **14** *A diagrammatic model of the fundamental process employed in the development of predictive models.*

model for simulating parking lot operations, or it can be one selected from the store of "ready-made" models as in the following example. An engineer has selected a model from a book, hypothesizing that it will satisfactorily predict the deflection (d) of beams he has specified for a structure (margin, next page). Here are some important assumptions that go with this model:

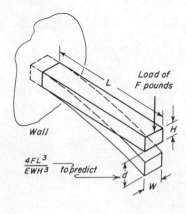

$$\frac{4FL^3}{EWH^3} \quad \text{to predict}$$

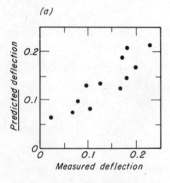

(a)

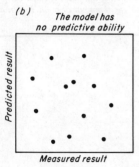

(b)

The model has no predictive ability

- The load F is applied at a single point.
- The material from which the beam is made is homogeneous throughout.
- The load F is applied gradually, not by impact.
- The beam is held perfectly rigid at its supported end.
- The weight of the beam is negligible.

It is obvious from these assumptions that there are a number of discrepancies between model and actual situation. (The model owes its relative simplicity to these assumptions.) The discrepancies are cause for concern; the engineer wonders about the accuracy of this model's predictions and its adequacy for the applications that he has in mind. Before he uses the model in his work, he intends to evaluate its predictions.

2. Therefore his next step is to obtain some observations from the real world with which the model's predictions can be compared. In this instance the observations of what actually occurs are obtained from laboratory experiments. Different loads are applied to beams of different L, E, W, and H values, and the resulting beam deflections are *measured*. Then, using these same L, E, W, and H values, the engineer *computes* the deflections from the model. Thus, for each set of conditions tested, he has a predicted deflection and an observed deflection, and he can plot predicted versus observed results. The greater the spread of points on a plot like (*a*), that is, the weaker the correlation between predicted and observed values, the poorer is the model's predictive ability, (see *b*).

3. Now, the engineer must interpret these results and make a decision. He has two major choices: (1) accept the model as is and use it; (2) repeat this three-stage process, starting with an attempt to refine the mathematical model originally proposed or with a different model, mathematical or otherwise.

This decision depends heavily on the situation in which the predictions are to be used. When the costs of errors are very high, especially when life and limb are at stake, the predictions must be quite accurate and precise. In other situations fairly rough predictions are adequate. Hence the adequacy of a model cannot be evaluated independently of the particular use to which it is to be put. Therefore you should not attempt to judge the seriousness of the dispersed points in graph (a) unless you have specific information about the manner in which the predictions are to be used.

You cannot expect the perfect predictive ability demonstrated in (c), a hypothetical plot, for several reasons. The points will never fall on a straight line because assumptions will inevitably be violated to *some* degree whenever a model is applied. Errorless predictions are unattainable. Furthermore, there is some error in the measurement of the values substituted into the model, for example F and E. Note too that predicted deflection is compared with *measured* (i.e., observed), not actual, deflection. Thus some of this dispersion of points is attributable to errors in measurement; don't blame it all on the model.

Obviously, as more time is spent in attempting to refine a model, its development cost continues to mount. Engineers *are* concerned about such costs. Under the usual pressure to keep costs low and to produce results as soon as possible, naturally you will invest no more time in the refinement of your models than is necessary for the purposes at hand. This whole matter comes under the heading of optimizing your problem solving methods, the topic for Chapter 13.

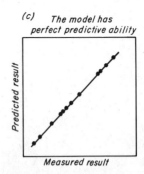

(c) The model has perfect predictive ability

Predicted result

Measured result

Summary Many of the models I have mentioned—mathematical equations, 3-D models, graphs, diagrams, etc.—are old hat to you. In these instances what is probably new to you is the point of view, namely, that these things have something in common as expressed in the term *model*. However, the situation might well be different with simulation; in this case you probably have learned something new. I am thinking mainly of digital simulation; it is surprisingly useful and amazingly simple. Incidentally, there is nothing old about digital simulation; obviously it is a relatively new technique since digital computers haven't been around that long. So if you know something about this technique, or, better yet, can prepare digital simulations, you will be one up on many of the elders in your field.

Exercises

1 *Survey other textbooks you are using (or have used) and identify three examples of each of the following types of models: iconic, graphic, diagrammatic, and mathematical. A brief verbal description of the illustration will suffice. Identify the book and the page on which each example is found.*

2 *On what assumptions is the equation of state of an ideal gas (page 51) based? Since real gases do not satisfy some of these assumptions, are the predictions provided by this model usable? Explain.*

3 *Choose a predictive model from another of your textbooks and identify all the assumptions that go with it. Then indicate the assumptions for which there probably is a significant deviation from the real world.*

4 *Develop a model for predicting the amount of time you require for various sizes and types of reading assignments. Most of the predictions provided by this model should be within plus or minus 20% of your actual reading time. Explain why the actual time and the predicted time for a given assignment very rarely agree.*

5 *A large railroad is planning to consolidate several of its freight classification yards into a large new yard. Freight trains arriving at the proposed yard are to be classified by the common "humping" process illustrated in Figure 15. The schedule of incoming trains has been tentatively selected as shown in Table 1. Before the new yard is constructed, the designer must determine whether one classification hump can accommodate this schedule. He must also predict the average delay of incoming trains and the amount of holding trackage that should be provided to store trains awaiting classification. To predict what will result, assuming the schedule given, the engineer is going to simulate operation of the classification process.*

For this purpose he has gathered certain data from routine records kept by the company and from actual observation of existing yards. This information is summarized in Tables 2 and 3 and Figure 16.

You are to perform a digital simulation of the classification process. In doing so:

(a) *Set up a tabulation sheet. Figure 17 gives you a start.*

Table 1
Schedule of incoming trains

12:17 A.M.	4:34 P.M.
12:49	6:38
1:28	7:10
1:36	7:59
1:51	8:22
2:20	9:15
2:35	9:55
2:48	10:04
3:19	10:49
3:49	11.03
4:30	11:42
5:57	11:51
9:11	

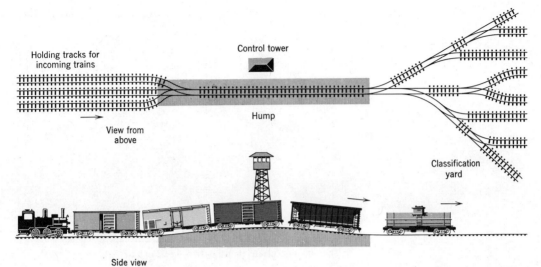

Holding tracks for
incoming trains

Control tower

Hump

View from
above

Classification
yard

Side view

Figure **15** *A simplified view of the classification system, showing the holding tracks where incoming trains await classification, the single-channel hump over which all cars must pass, and the multitrack yard in which new trains are made up.*

Figure **16** *Train-size data for a sample of 100 incoming trains. This type of graphical model, a frequency histogram, is a very useful means of summarizing a large number of numerical values. It indicates, for example, that 7 of the 100 trains had 116 cars.*

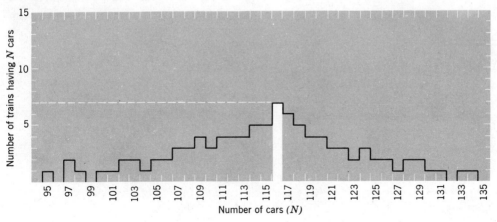

75

Table **2** *Train arrival data—deviations (in minutes) from scheduled arrival time*

+2*	+13	+2	+6	+9
+8	+11	+4	+5	+13
+12	+8	+6	+2	+21
−6	+5	+9	+15	+17
0	+1	+5	+7	+9
+4	+14	+10	+10	+3
+6	−3	+12	+8	+5
−2	+6	+17	+12	−4
+1	+2	+6	+5	+8
+3	+11	+3	+2	+6
+10	+5	−1	−7	+19
+22	+1	−8	+6	+10
+4	+16	+3	+7	−2

* The plus sign means that the train was 2 minutes late.

Table **3** *Data collected on classification time*

Number of Cars in Train	Minutes to Classify	Number of Cars in Train	Minutes to Classify
113	35.4	114	34.0
121	33.9	100	29.7
129	35.7	117	32.3
108	33.9	127	40.1
117	34.9	116	36.7
123	34.2	135	41.1
109	36.0	121	34.1
128	37.9	124	36.1
135	36.9	109	34.8
124	39.4	122	37.8
118	34.2	127	36.2
112	37.4	110	32.8

(b) *For each type of data from which you must make random selections, use a different method. Two of these Monte Carlo methods are described in this chapter; a third is given in Appendix B.*

(c) *Demonstrate your procedure by simulating at least a dozen train classifications, starting with the 4:34 P.M. train.*

(d) *Carefully explain by words, diagrams, etc., the procedure you have set up.*

(e) *Describe the assumptions made by your model and indicate those you suspect may have to be removed if your model's predictions are not satisfactory.*

(f) *Assume that your simulation model has synthesized four days of experience with the proposed schedule, and that the times trains had to wait for classification were accumulated, yielding the result shown here. What are some things you could point out to the railroad's management on the basis of these results?*

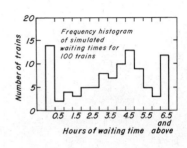

(g) *Prepare a flow diagram of a procedure a digital computer can follow in executing this simulation. (See Appendix B.)*

Fig. 5-17

TABULATION SHEET FOR SIMULATION OF CAR CLASSIFICATION							
Times	Source of Time Values	Trains					
T_s (scheduled arrival time)	Proposed schedule	4:34 P.M.	6:38 P.M.	7:10 P.M.	7:59 P.M.	8:22 P.M.	
D (deviation from scheduled time)							
T_a (actual arrival time)	$T_a = T_s + D$						
T_c (time at which classification begins)							
Wait time							

6 *Develop a digital simulation model of the game of baseball.*

6

OPTIMIZATION

WHEN you adjust the focusing knob of binoculars you are engaged in a process the engineer calls optimization. In such instances there is a criterion (e.g., image sharpness) that depends on a second variable (e.g., distance between the lenses), which you manipulate in order to maximize the criterion. The *optimum* value of the manipulated variable is the one that yields this maximum. Thus the optimum lens adjustment is the one that produces the sharpest image.

This is a common situation. In each case there is (a) a dependent variable, called the *criterion,* which is influenced by a *manipulated variable,* and (b) a value of the manipulated variable for which the dependent variable is a maximum, called the *optimum value.* The manipulated variable in each of these cases has an optimum value with respect to the criterion indicated:

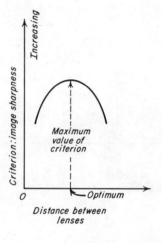

Manipulated Variable	Criterion
Fuel injection rate	Engine efficiency
Room temperature	Body comfort
Price charged	Total revenue
Rate of work	Total work accomplished

The concept of optimum is an important one in engineering. There is an optimum solution to almost every problem. In fact, each specific characteristic of a solution has an optimum value. For example, there is an optimum size and shape for a coffee-pot handle with respect to ease of handling, an optimum mix of weapons a nation should stock, an optimum process for refining petroleum, and an optimum mix of ingredients with respect to the strength of concrete. Thus the concept of an optimum permeates most aspects of an engineer's work. It guides his actions and decisions; it serves as a goal both in the solutions that he produces and in the manner in which he arrives at them.

Optimization is the process of seeking the optimum value, condition, or solution. Unfortunately, in most engineering problems optimization is much more complex and time consuming than in the binocular-focusing case. This is so primarily because of the numerous conflicting criteria.

Conflicting criteria, although you may not have called them that, are certainly not new to you. You have dealt with them if you have tried to tune in a television program when the clearest picture and the clearest sound do not occur at the same setting of the dial. The optimum settings for picture clarity (O_p) and sound clarity (O_s) do not coincide; the setting that is optimum with respect to picture clarity is *suboptimum* for sound, and vice versa. In this situation you must compromise between two conflicting criteria, picture clarity and sound clarity—conflicting in the sense that, as you try to improve the situation with respect to one criterion, you are likely to make it worse with respect to the other. Where between O_p and O_s you set the dial depends on the relative importance of picture clarity and sound clarity *to you*. Since different persons attach different degrees of importance to these two criteria, their settings will not coincide.

This situation—conflicting criteria and the need to find a compromise between them—abounds in engineering problems. Usually, however, the conflict is among many criteria. Take the engineer who is developing a machine for harvesting fruit. Among the criteria he must consider are the speed with which the machine picks fruit, the safety to persons near it, the degree to which it damages fruit, and the cost (of construction, operation, and maintenance). He cannot specify a machine that is the ultimate in speed unless cost is no object and no one cares about safety and fruit damage. But the prospective purchasers of such a machine *do* care. Hence he must sacrifice some speed to gain a reduction in cost, fruit damage, and hazard. And too, he could design a machine that does not damage fruit, but it would probably be so slow and so expensive that no one would buy it. Similarly, the machine could be made almost perfectly safe, but only at a cost few potential buyers would consider paying, and so on for all criteria. Consequently, the engineer alters his design until he achieves what he

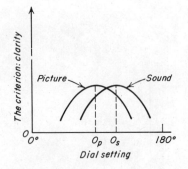

Dial setting

believes is the optimum balance between the conflicting criteria.

Optimization *is* a compromising process, and it becomes very complicated when there are more than two conflicting criteria or when they cannot be put into numerical terms. Yet it is a necessary process if the final solution is to be optimum. In a relatively simple case it proceeds somewhat as follows. To determine the best balance between harvesting speed and lack of damage to fruit, the engineer must know the relationship between these two criteria. On the basis of previous experience and some direct experimentation he estimates that damage depends on speed approximately as shown here. With the help of this model, the engineer can predict the extent to which fruit damage can be reduced by a given sacrifice in speed, or what a given increase in speed will cost in terms of damage. In the course of such deliberations the engineer is determining how much speed to trade for reduced damage in order to achieve the optimum compromise. This "giving and taking" between criteria, in order to reach the best balance, is appropriately referred to as the *trade-off process.*

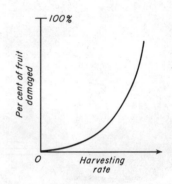

Trade-offs   The trade-off process is all too familiar to the experienced engineer. Almost all of his decisions involve trade-offs, and many of them are difficult to make. Some examples will be helpful.

Making a ballistic missile more destructive seems like a simple matter: just increase the size of the bomb it carries. But this increase adds weight and the missile is already carrying its maximum payload; hence to compensate for the greater bomb load a sacrifice must be made in the missile's range (by reducing its fuel supply) or in its ability to land on target (by reducing the guidance equipment it carries) or in its ability to penetrate enemy defenses (by reducing the penetration aids it carries). Trade-offs must be made between these criteria until the optimum compromise is found.

In laying a telephone cable along the ocean floor, the chances of breakage are minimized if plenty of slack is allowed. But slack increases the amount of cable needed. If the cable is kept relatively taut, the consequences are reversed. A trade-off must be made.

It generally follows that cheap engines have higher op-

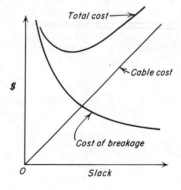

erating costs (fuel, maintenance, repair) than expensive ones. Therefore the purchaser often finds it profitable to make some sacrifice in purchase price to achieve a gain in operating cost. This type of trade-off situation is familiar to you.

You can visualize some of the trade-offs that must be made in the design of a high-speed mass transportation system, where such criteria as speed, safety, comfort, privacy of accommodations, load-carrying capacity, convenience, and construction cost are important.

Figure 1 This is the reflector of an antenna used to track and communicate with space vehicles anywhere in the solar system. Some significant trade-offs were made in its design. For example, the antenna's range can be increased by enlarging the reflector, but this makes it more difficult to aim especially due to increased vibration caused by winds. The designers had some difficult compromising to do in order to achieve a satisfactory balance between range and "aimability." (Courtesy of NASA)

Value Decisions The engineer cannot arrive at the optimum trade-offs between criteria until he knows the relative importance of each (recall the television tuning example). Knowing the relationship between harvesting speed and damage to fruit is of limited help until the engineer learns the relative importance of different speeds and degrees of damage to potential users of the machine. Perhaps most users attach considerable value to high harvesting speed and are willing to tolerate a rather large percentage of damaged fruit in order to obtain this speed. Once the engineer knows the relative values attached to criteria, he can determine the trade-offs that must be made in order to optimize the overall design.

The assigning of a relative value (importance, weight) to a criterion is commonly called a *value decision*. In engineering value decisions are difficult to make, partly because the engineer must anticipate the value attached to a given criterion by others, often a large group at that. For example, designers of the mechanical toothbrush had to determine whether a substantial proportion of the potential users would be willing to pay X dollars more to avoid the inconvenience and hazard of a direct cord connection to a 115-volt electrical outlet.

Value decisions are never more difficult than when they involve human life. The designer of a highway system must consider criteria such as construction cost, capacity, safety, durability, and maintainability. Construction cost and safety are conflicting criteria. He could specify a virtually impenetrable dividing wall for a four-lane dual highway under design, thereby adding $750,000 to total construction cost and reducing fatal accidents by two thirds. Is the added safety worth the cost? Some taxpayers would approve the extra expenditure if it promised to save only one life. Others would react differently, as indicated by the absence of such barriers on many four-lane highways.

This "dollars versus life" dilemma is only one of many instances in which criteria cannot feasibly be quantified in common units of measure. Dollars are convenient units, but many criteria cannot be expressed in monetary terms —safety being a conspicuous example. In fact, many criteria, like safety, are difficult to measure in *any* units. Yet safety is a criterion to be reckoned with in most engineering problems. The inevitable result is that an engineer has some very trying value decisions to make.

Optimization Procedure A traffic engineer is making a study with the objective of maximizing the number of vehicles that can pass through New York City's heavily traveled tunnels. No, this is not simply a matter of speeding up the vehicles. You know and so does he that drivers increase the spacing between their vehicles as their speed increases. In fact, he has data (margin) for a large sample of drivers which indicate that spacing increases at a faster rate than speed. Therefore, if drivers are encouraged to move faster in an attempt to get more through a tunnel per hour, the increased spacing will have the opposite effect. This causes the engineer to suspect that some average ve-

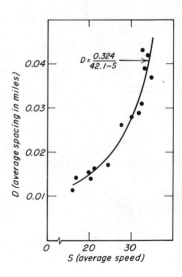

$$D = \frac{0.324}{42.1 - S} \qquad (a)$$

$C = $ vehicles per hour

$$C = \frac{S}{D} = \frac{miles/hour}{miles/vehicle} \qquad (b)$$

Combining equations a and b:

$$C = \frac{S}{\dfrac{0.324}{42.1 - S}}$$

$$\boxed{C = 130.0S - 3.1S^2} \qquad (c)$$

↳Criterion function

Figure 2

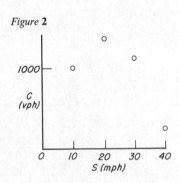

Figure 3

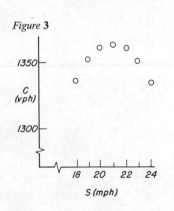

hicle speed is optimum, that is, maximizes the number of vehicles handled per hour. He proceeds to explore this notion in the following manner.

From his spacing versus speed data he derives mathematical model (a). Here D is the average spacing between vehicles and S is the average speed. He then converts this model into equation (c) which describes the effect of S (the manipulated variable) on the number of vehicles passing through the tunnel per hour (the criterion). Mathematical model (c) is called a *criterion function;* it describes in what way the criterion is a function of one or more manipulated variables.

To determine what value of S maximizes C (i.e., what speed will get the most vehicles through the tunnel per interval of time), the engineer substitutes a trial series of speed values into equation (c) and computes the resulting values of C. The results appear in Table 1. S values in multiples of 10 are used to determine the general location of the maximum C, and then a "localized region of interest" around 20 mph is explored by substituting S values in increments of 1 mph. The engineer's conclusion: 21 mph is the optimum speed with respect to the vehicles-per-hour criterion.

Table 1

Trial values of S (miles per hour)	Resulting values of C (vehicles per hour)	
10	990	First set
20	1360	of trial
30	1110	values
40	240	(Figure 2)
18	1336	
19	1351	Succeeding
21	1363	trial
22	1360	values
23	1350	(Figure 3)
24	1334	

The same conclusion can be arrived at by applying elementary calculus. The criterion function, equation (c), is differentiated with respect to S:

$$\frac{dC}{dS} = 130.0 - 2(3.1)S = 130.0 - 6.2S \qquad (d)$$

Note that dC/dS is the rate at which C changes as S changes, and is the slope of the curve in Figure 4. It is apparent that this slope is different at different values of S. At the peak of the curve where C is a maximum this slope is zero. Therefore the optimum speed is the value of S for which

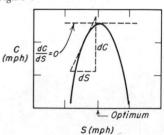

Figure **4**

$$\frac{dC}{dS} = 0 \qquad \text{(e)}$$

But

$$\frac{dC}{dS} = 130.0 - 6.2S, \qquad \text{(f)}$$

Therefore

$$0 = 130.0 - 6.2S$$
$$S = 21 \qquad \text{(g)}$$

Thus the optimum value of S is 21 mph.*

These alternative means of arriving at the optimum vehicle speed illustrate two basically different methods of seeking an optimum solution. One of these is referred to as the *iterative* (or *numerical*) *method*. In general it goes like this, using Table 1 to illustrate.

1. The engineer assumes for the manipulated variable a value that in his judgment is optimum, or he may choose a range of values that look promising. (Four vehicle speeds were selected.)
2. Then he predicts what effect the assumed value(s) will have on the criterion. (Equation (c) was used for this purpose.)
3. Benefiting from this exploratory probe, he selects one or more new values of the manipulated variable. Presumably the new value(s) will be closer to the optimum than his original guess. (The values of C for the first set of S values in Table 1 indicate that the optimum S must be in the vicinity of 20 mph. Therefore a value of $S = 18$ was selected for the next trial.)
4. The engineer predicts the effect on the criterion as before, and accordingly selects a new trial value if necessary. (C for 18 mph is less than that for 20 mph, so $S = 19$ was selected for the next trial.)

This cycle of events is continued until the engineer is satisfied with his estimate of the optimum value. (Thus values of $S = 21$, then $S = 22$, etc., were tried until the engineer

* This illustration is based on studies made by the Port of New York Authority, as reported in "The Influence of Vehicular Speed and Spacing on Tunnel Capacity," Edward S. Olcott, JOURNAL OF THE OPERATIONS RESEARCH SOCIETY OF AMERICA, Vol. 3, No. 2, May 1955.

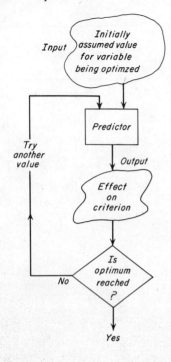

$$C = f(V)$$

where

C is the criterion and V is the manipulated variable

$$C = f(V_1, V_2, V_3, \ldots V_N)$$

was convinced that the maximum *C* is at 21 mph.) Graphs like Figures 2 and 3 are useful supplements to or substitutes for a tabular summary like Table 1.

The iterative method of optimization is basically an accelerated learning process. Through a series of successive approximations the engineer gradually "closes in" on the optimum value of a manipulated variable. The function of the "predictor" is to tell the engineer what effect an exploratory change in the variable being optimized will have on the criterion, so that he can learn from this and move closer to the optimum in the next trial. The "predictor" can be a mathematical or simulation model, a real-life experiment on a prototype, or the engineer's judgment. The mathematical model on page 68 for predicting the average waiting time of barges was being used for this purpose; likewise for the simulation models described on page 57.

The other of the two basic techniques for locating an optimum is referred to as the *analytical method*. In this case a mathematical model yields the optimum value directly, as illustrated by the second method of arriving at the optimum vehicle speed in tunnels. Often the model is derived through the use of calculus, generally proceeding as follows.

1. A criterion function, e.g., equation (c) is developed. Although the criterion is often measured in dollars, it need not be. It can be number of passengers carried, pounds, efficiency, etc.
2. Then, by differential calculus or other means, the criterion function is converted to a form that yields the optimum value of the manipulated variable directly.

In either the iterative or the analytical method the criterion function can contain more than one variable to be optimized.

Surely you wonder why anyone would use the iterative procedure considering the directness of the analytical method. It's a good question and there is a good reason, namely, that in many cases the analytical method is much too difficult mathematically and the iterative method is the only practical way. Furthermore, in many instances the engineer wishes to know how *sensitive* the criterion is to deviations of a variable from its optimum value, and the iterative method ordinarily provides this information conveniently. Note from Table 1 that very little is lost in terms of vehicles per hour if the average speed deviates slightly from the 21 mph optimum. If the average speed is 25 mph, there is only a 4% loss in vehicles per hour, which isn't bad. This information is useful because it would be difficult to attempt to control the average speed to 21 mph, and so the engineer is interested in learning what sacrifice in vehicles per hour would be made if he attempted to control speed to some higher value. An investigation of this kind, to learn the consequences of not setting a variable at its optimum value, is a *sensitivity analysis.*

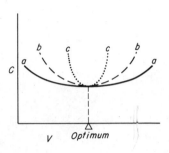

The iterative and analytical optimization methods illustrated above cannot be relied on as much as the engineer would like. Because of a preponderance of unquantifiable criteria, such as esthetic appeal, or because of a large number of variables or of lack of time, engineers often must rely on procedures that are less formal, less quantitative, and less objective than those illustrated. Usually they use a variety of different methods in their optimizing efforts, the particular combination of procedures varying considerably from problem to problem (so much so that it is foolish for an author to attempt to generalize any more than I have).

Optimum as a goal An important goal in an engineer's endeavors is the "optimum." In design he searches for the optimum solution to a problem and strives to do so by optimum means. Notice, however, that I did not say he attains; I said he searches. Although the optimum solution is almost always an objective, it is certainly not always a realization. Many real-world problems are too complex for an optimum solution to be found in a reasonable period of time. In many instances the amount of time required would be longer than the life of the problem. Almost invariably many other problems await the engineer's attention, and it often becomes more profitable for him to shift to one of these than to continue searching until the optimum is achieved for the original problem. Thus, in practice, it is usually a matter of progressing toward the optimum, continually seeking successively better solutions until it becomes more profitable to spend the effort elsewhere.

Exercises

1   For any two of the following devices, structures, or processes, describe what you believe are important criteria that the designers must have considered. Identify criteria that appear to be conflicting and, therefore, those between which trade-offs probably had to be made.

   (a)   An interchange between two intersecting dual highways.

   (b)   A giant ocean-going luxury liner.

   (c)   An artificial hand.

   (d)   An automobile.

   (e)   A large machine for making electric light bulbs.

   (f)   A complete factory for manufacturing refrigerators.

   (g)   Any one of the case studies described in Chapter 2.

2   Cite ten familiar situations in which there is obviously an optimum value for some variable with respect to a stated criterion. (Example: there is an optimum reading speed with respect to total knowledge assimilated.)

7
COMPUTATION

YOU have been developing your ability to compute since your preschool days, so this is a familiar skill. However, your computation so far has been mainly mental, usually aided by pencil and paper, which for short jobs is the logical method to use. But many computational tasks in engineering are of a size you have yet to see; hence you had better be prepared with some faster methods. You wouldn't want to be equipped only with pick and shovel for jobs calling for a bulldozer. Therefore in college you will extend your computational skills to include at least the slide rule and digital computer and probably the analog computer and desk calculator. Thus you will be equipped with an array of computational tools suitable for jobs ranging from the simple one-shot calculation to the gigantic task involving thousands of routine, repetitive operations.

The remainder of this chapter is a survey of the ways in which engineers use digital computers. Do not conclude that other computational subjects are unimportant. However, there are numerous books on such topics as computer programming and slide rule operation. Therefore in the limited space available I am covering what is lacking in current literature, and that is an introductory overview of how computers are applied in engineering.

Engineers use computers in so many ways that it would take more than the pages of this chapter simply to name them. Fortunately, most of these uses can be classified into a reasonable number of types, as summarized in Figure 1. In general, the computer is used as a tool (like the transit and slide rule) in arriving *at* solutions to problems. Engineers also use computers *in* many of their solutions, for example, as part of the spacecraft guidance system. I will illustrate.

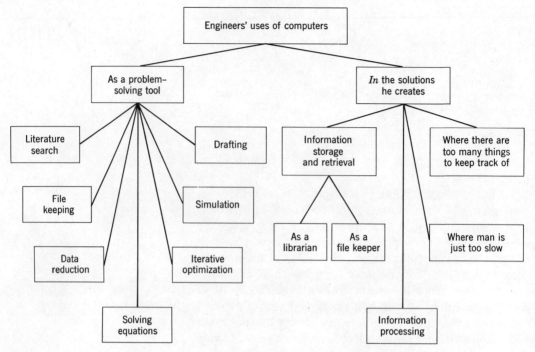

Figure **1**   *The "big picture" for this chapter.*

Use of Computers in Design

The computer is dramatically affecting the practice of engineering. It is fast becoming an indispensable problem-solving tool, aiding the engineer in a variety of ways.

Literature Search   An engineer is designing an integrated information system for a university. This computer-based system will process grades, prepare schedules, handle accounts, centralize student records, do some instructing, and perform a multitude of other useful functions. The engineer surmises that systems to handle various parts of this problem have been developed by individuals at many institutions and companies, but how does he learn who the people are and what they have developed? He could easily shoot several weeks in tracking down reports of work done on this sort of thing elsewhere, through library searches, reading, talking to people, corresponding, telephoning, and traveling. And after all this effort he would have missed some worthwhile systems others have worked out, simply because there are so many scattered references and because a really exhaustive search would be too expensive and time-consuming.

This is the way it is for most engineering problems. The literature search tasks are staggering. Ordinarily the engineer conducts a reasonable search and then proceeds to solve the problem at hand, accepting the risk that he is duplicating the work of someone who has already solved part or all of the same problem. Under the circumstances it is uneconomical for him to do otherwise.

But the computer is coming to the rescue, enabling the engineer to conduct his search like this. He lists keywords, the kind of words he would keep in mind if he were conducting a search in the library. In this case he would have two lists, including such words as these. He submits these to a computer center to be used for search of the files on computer applications. These files consist of magnetic tapes containing abstracts that describe every computer application made public. The computer goes through these tapes, and for every abstract that contains at least one word from list 1 and at least one word from list 2 it prints the abstract and the information the engineer needs to get the details. Within the half hour he has a much more comprehensive collection of references than he could possibly afford to gather by customary methods. (Note that in this example the engineer is using the computer in his solution *and* as an aid in arriving at that solution.)

Such literature search systems are not in widespread use, but you can see why they are under development. They will minimize the costly duplication of problem-solving efforts, save engineers from having to sift through "mountains" of books and technical reports, and do the job quickly, cheaply, and exhaustively.

Keyword List 1	Keyword List 2
College	Admissions
School	Billing
University	File keeping
Student	Inventory
	Record keeping
	Registration
	Scheduling

Records, Records, Records   In any engineering department there are extensive files of drawings describing previous designs created by the department, right down to the details of each gear, cam, and shaft. An engineer designing a motor is currently considering its bearings. He knows that bearings have been designed by others in the department and also that they can be purchased from dozens of suppliers. Before he goes to the trouble of designing them himself, he wants to check what's already available in the hope that something will be suitable. But he doesn't want to spend hours thumbing through cabinets full of engineering drawings and through the catalogs of bearings suppliers. Yet this is exactly what was necessary before the computer came to the rescue. Now a system is available that asks him only to specify roughly what type of component he is looking for, and then within minutes tells him what his company makes and what can be purchased to satisfy his needs. This system serves as the engineering department's general filekeeper, storing most of the information formerly kept in file drawers, log books, catalogs, and drawings cabinets. It answers most inquiries in seconds if not fractions of seconds. At many companies such systems remain to be developed, but again it's only a matter of time.

Data Reduction   Engineers often have large volumes of data to be reduced to useful form, for example, hundreds of measurements from an experiment. Calculating averages and measures of variability, curve fitting, statistical tests, etc., are usually time consuming and tedious if done by hand. The computer eats this stuff up.

Solving Equations   Most common mathematical operations can be executed by the computer, and thank goodness. Some of them, like solving simultaneous equations of many unknowns, take hours or days to do by hand. Unlike the equations ordinarily found in mathematics books, those encountered in practice are often very messy to solve. The $y = ax^b$ of the textbook is likely to be $y = 1.378x^{2.3}$ in practice. As a result, computation can become very time-consuming. Try finding the value of x that balances the equation $x = 1.31e^{0.27x}$ to within 1%. This will take some time by pencil and paper methods. You can see why the computer is so important to engineers as an equation solver.

Iterative Optimization   Recall from page 84 that this procedure can become time-consuming if done by hand. In fact, optimization by formal iterative methods was not commonly attempted before computers became available. Now engineers can take full advantage of this powerful technique.

Simulation   A fast-spreading technique is digital simulation by computer (Appendix B). It is possible to conduct

(a)

Figure 2a This cathode ray tube (CRT) enables a computer to display alphanumeric information. The user communicates to the computer via the typewriter keyboard or the light pen he holds. For instance, this physician can indicate which of the tests listed on the screen he wants performed on a patient by pointing to them with the pen. (Courtesy of the Burroughs Corporation.)

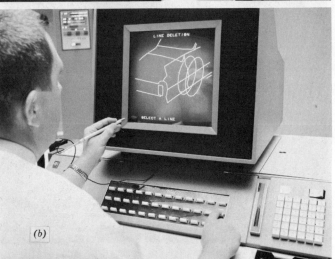

(b)

Figure 2b This combination of computer, CRT, man, and light pen can accomplish remarkable things. Man can draw on the face of the tube with the pen and the computer can remember what is drawn and reproduce it on the screen whenever instructed to. Even better, the computer can manipulate these drawings. For example, it can show you the object from different perspectives, rotate it, or enlarge it, as instructed via the pen and the keyboard. In this particular view, the operator has pressed the "line deletion" button which enables him to point to any line and have it erased from the screen and the computer's memory. (Courtesy of General Motors Research Laboratories.)

experiments on computers quickly and economically, with complete control of the experiment. (You know why engineers turn to the computer to carry out digital simulations, especially if you have done one by hand.)

The "Computer Draftsman" Auxiliary equipment now enables you and the computer to communicate graphically.

Figure 3   These before and after photographs of the screen pictured in Figure 2b illustrate some of this system's capabilities. When the engineer finishes his work and wants to make a permanent record of what appears on the screen, he can do so by instructing the system to store it, in which case it can be called back to the screen at any time. He can also have the computer instruct an X-Y plotting machine which will prepare an inked drawing, or he can obtain a permanent copy by photographic means. (Courtesy of General Motors Research Laboratories.)

Give the
computer these:

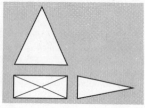

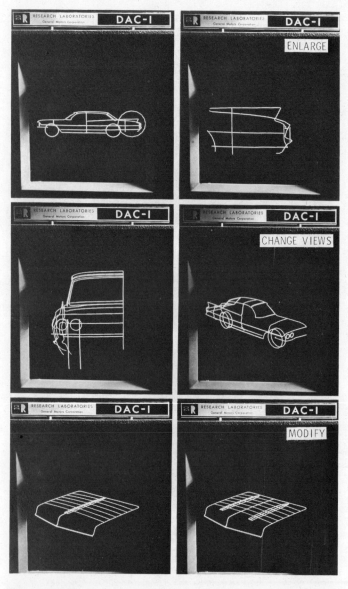

and
It will give you

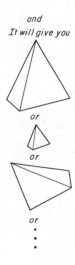

This remarkable and significant development is just beginning to affect the practice of engineering, but within a few years the impact of *computer graphics* will be great. You can see from Figures 2–4 that this system has real potential.

This list of the ways in which engineers use computers in their day-to-day work is not exhaustive. It is a sampling, to give you some idea of the usefulness of these machines.

Use of Computers in Engineers' Creations

In general, if an engineer incorporates a computer *in* the system he is designing, he does so because his solution requires at least one of the following:

1. An economical means of *storing* information.
2. An economical means of *processing* information.
3. A means of handling information at speeds *only* the computer is capable of.
4. A means of *keeping track* of many concurrently changing, interacting events or variables, in situations where the computer is the best if not the only way of doing this.

Some elaboration is called for.

Information Storage and Retrieval The preservation of knowledge so that it can be relocated without unreasonable time and expense is a critical problem in most fields of human endeavor. The story is the same in medicine, law, business, education, and government, as well as engineering. Brains, books, and file drawers—the familiar means of storing information—are becoming hopelessly inadequate in many cases. But the computer has a memory that is remarkably reliable, large, and fast. And here there is real hope.

Mankind has two major types of information storage problems. One concerns the storage of general knowledge (e.g., all scientific knowledge), which at present is typically found in libraries. The other type involves private information (e.g., an insurance company's files on its policyholders), which now is typically stored in the office files of business corporations, government agencies, and institutions. Computers are very good at both types of task; engineers will be incorporating them frequently in solutions to information storage problems, primarily when large bulks of information must be searched frequently. Here are some sample applications.

Storing and Retrieving General Knowledge by Computer What engineers are developing for themselves in this respect (page 91) they can set up for others. For example, representatives of a state government requested an engineering firm to develop a more effective system for storing and searching the state's legal information, which includes

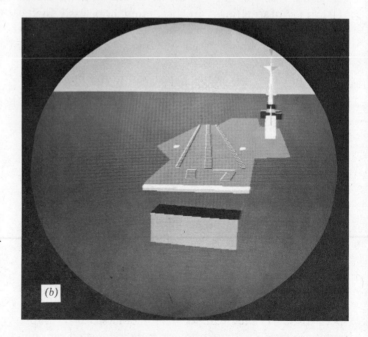

Figure 4 Given a series of numbers which locate the corners of an object as X-Y-Z coordinates, this computer system will prepare television pictures of that object at any perspective and to any scale. Since it can change views up to 30 times a second, it can give the illusion of movement—of you with respect to the object or the object with respect to you or both. Views like those shown here can be made to change continuously in response to your movement of typical aircraft controls, thus allowing you to practice landings on this carrier—in color at that. Obviously this system is useful for simulation. (Courtesy of the General Electric Company.)

(c)

all statutes and all court decisions. The heart of the solution to this problem is a computer, with supplementary equipment and specially prepared computer programs.

Now, for example, lawyers, judges, and legislators can use this system to search the state's statutes in order to isolate all laws pertaining to a given subject, such as narcotics. For this purpose a specially prepared magnetic tape (called a concordance tape) is used. It contains all significant words appearing in the statutes and identifies every statute in which a particular word is used. Suppose a lawyer wishes to know what statutes have something to say about the *education* of *handicapped children*. He prepares keyword lists and submits them to the computer via punched cards. The computer processes the concordance tape and prints the numbers of all statutes that contain the words *education, handicapped,* and *children* (or any synonyms of these words). In fact, it will print the relevant statutes themselves if so instructed.

Storing and Retrieving Private Information Engineers are being called on frequently to help hospitals solve their cost, labor shortage, and information-handling problems. The last of these is probably the most critical. In a hospital much time is spent in recording, filing, exchanging, and

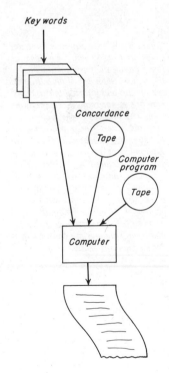

Key words

Concordance
Tape

Computer program

Tape

Computer

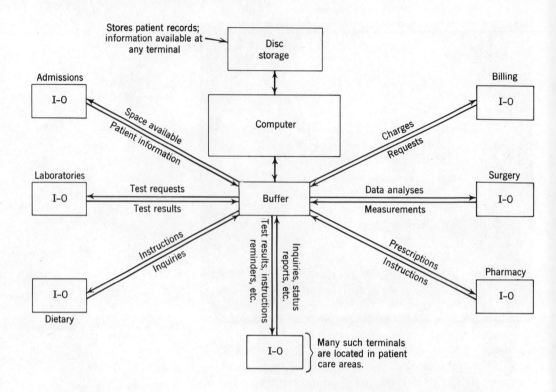

Stores patient records; information available at any terminal → Disc storage

Admissions
I-O

Space available
Patient information

Computer

Charges
Requests

Billing
I-O

Laboratories
I-O

Test requests
Test results

Buffer

Data analyses
Measurements

Surgery
I-O

Instructions
Inquiries

Test results, instructions reminders, etc.

Inquiries, status reports, etc.

Prescriptions
Instructions

Pharmacy
I-O

I-O
Dietary

I-O

Many such terminals are located in patient care areas.

Figure 5 A "computerized" master record-keeping system for a hospital. From any of the terminals located throughout the hospital an authorized person can add information to a patient's file or request information from it. Examples of the types of data that this system stores, processes, and transfers are indicated by the arrows. Equipment at a typical input-output station is pictured by the inset. This computer can also serve as the heart of a patient-monitoring system, which alerts hospital personnel when a change in the pulse, temperature, etc., of a critically ill person warrants attention. (Photograph courtesy of the Burroughs Corporation.)

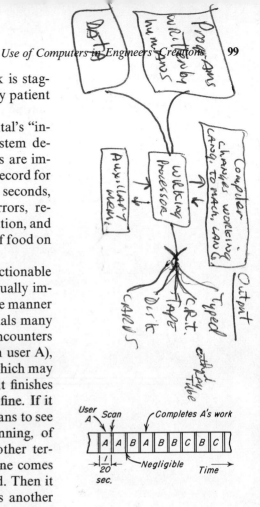

looking up information. The record-keeping task is staggering. Thousands of square feet are consumed by patient records.

A team of engineers called on to solve a hospital's "information problem" specified the computer system described in Figure 5. The possibilities and benefits are impressive. This system keeps an up-to-the-minute record for each patient, answers inquiries within a few seconds, checks prescriptions for inconsistencies and errors, reminds nurses when it is time to administer medication, and tells the kitchen how many pounds of each type of food on the menu will be needed for the next meal.

You might wonder about the possibility of objectionable delays with so many potential users. This is virtually impossible in view of the computer's speed and of the manner in which the system operates. It scans all terminals many times a second in round-robin fashion. When it encounters a terminal at which there is a customer (call him user A), the computer devotes ½0 second to his request (which may be to file, look up, or process information). If it finishes user A's job in that interval, as is fairly likely, fine. If it doesn't, it puts his job on ice, so to speak, and scans to see whether there are other active terminals. Scanning, of course, takes negligible time. If there are no other terminals at which someone has a task, the machine comes back to user A and gives him another ½0 second. Then it scans again, but suppose that this time it finds another active terminal, at which user B has work for it. It gives user B ½0 second, scans, and finding no new active terminals it returns to user A. And so it goes. Even if most of the 30 terminals were active simultaneously, which is very unlikely, user A's task would be completed so quickly that he would get the impression that he had exclusive use of the computer, whereas actually he is sharing it with others.

This mode of operation, called *time sharing,* is a relatively recent development in the computer field. You will hear a lot about it and surely feel its impact. To believe that a system like this exists requires that you have a feel for the fantastic speed of the computer. Actually, it is serving its customers sequentially, but it switches so rapidly and completes its work so quickly that it appears to be serving them simultaneously.

The buffer (Figure 5) is an ingenious device without which time sharing would be impractical. User A spent 7

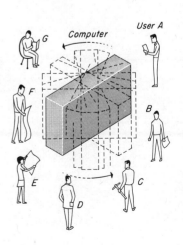

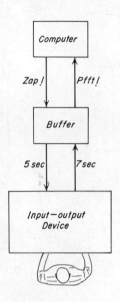

seconds typing his request for patient information. Fortunately, during that period his typewriter was not transmitting directly to the computer. His message went to the buffer, where it was stored until user A pressed the end-of-message key. Then, when the computer scanned to user A, the buffer fired his message into the computer in a few milliseconds. This is great. It frees the very fast and expensive computer from the relatively slow process of typing. The buffer, then, serves as a time compressor for incoming messages. It serves also as a time expander in the reverse direction, so that the 0.2 millisecond message it receives from the computer is fed to the typewriter at a speed the latter can follow. As you can see, this communication buffer between man and computer really pays off.

This application of the computer as the hospital's master file keeper exemplifies the use engineers have made of computers in systems for handling airline reservations, bank accounts, and inventory records. These systems are forerunners of the file-keeping systems that in the future will be common in insurance companies, motor vehicle bureaus, mail-order houses, school systems, and other situations where there are large-scale record-keeping requirements. The computer is destined to become society's "great file keeper." Incidentally, the majority of information storage-retrieval applications of the computer have yet to be made.

Economical Processing of Information   The engineer often incorporates a computer *in* his solution because it is the cheapest way of *processing* information in the situation at hand. In the preceding type of computer application information storage was primary, processing was incidental. Here the reverse is true. Probably most computer applications you are familiar with are of this type. This figures, for this is the most common use, and the one you usually read about.

When a very large number of repetitive operations must be performed on information, the computer is likely to be the cheapest way of doing the job. This is why an electric utility prepares its 120,000 bills each month by computer, why a college schedules its 30,000 students by computer, why a corporation prepares its 15,000-employee payroll by computer, and why the Census Bureau uses computers.

Repetitiveness is conspicuous in every instance. In each such case human beings can do the job although a small army might be required.

When Only the Computer Is Fast Enough   In the previous two types of computer application, man and machine compete for the task at hand and the computer wins out because it is more economical. But there is no choice here; man simply can't respond fast enough. Here is an example.

What may be an ICBM is detected by radar, apparently headed toward the United States. There is no time to spare. The nature of this object must be verified and a decision made concerning interception, all in a big hurry. Unfortunately, there is no time to fly out and have a look at the object for identification purposes. Those days are long since gone. The only practical way now is to track the object by radar and on the basis of this information compute its speed and probable destination, and then quickly see whether it can be accounted for on the basis of aircraft and spacecraft known to be in the area. On the basis of this information a choice must be made: forget it, wait, intercept. All of this must be accomplished in a few seconds; only a computer is capable of such speed. If an interceptor is to be launched, aiming it constitutes another challenging problem that only the computer can handle (Figure 6).

Conspicuous in this type of computer application is the speed with which things happen. Certain manufacturing

Figure 6   To intercept and kill an intruding ICBM, the path of the intruder must be predicted and an intercept point calculated, which is likely to be hundreds of miles from its present position. Hence the interceptor has a computed path it must follow to the point of kill. Unfortunately there are forces that tend to steer it off course. Therefore it is necessary to sense deviations from the intended path by radar, compute necessary corrections in course, and transmit these guidance instructions to the interceptor. This sequence of events—sense position, compute corrections, transmit—goes on continuously. This guidance system must react in microseconds in view of the missile's 18,000-mph speed. While all this is going on, radar continues to sense the intruder's movement and the computer "updates" its prediction of the intruder's path and target. Thus, the interceptor must be guided to a point of kill that itself is changing. And that's not all. An ICBM would no doubt be accompanied by clouds of harmless decoys. The real things must be sorted out in a hurry. Such matters complicate the whole process considerably. One thing is for sure: buildings full of men could not possibly make these computations in time.

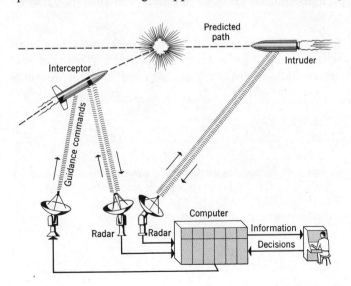

processes, like the rolling of steel slabs into flat sheets, operate at very high speeds, making it impossible for human beings to control them. Similarly, in an aircraft collision-avoidance system human reaction times are just not short enough. In these and an increasing number of other instances engineers must call on computers to do the job.

Too Many Things to Keep Track of   Suppose there are 30 planes in the air around an airport. How can a man keep track of them all? He can't! But a computer can—see Figure 7. For all practical purposes the computer is keeping track of all planes continuously and simultaneously. Actually, however, it is concentrating on one plane at a time, switching its attention from plane to plane and updating its knowledge of the whereabouts of each at more than satisfactory frequency. This high-speed scan process is of course characteristic of time-sharing computer systems.

Some manufacturing processes parallel the air traffic control situation in that they involve perhaps dozens of variables (temperatures, pressures, speeds, etc.) that are subject to frequent change and are interrelated. No human being or any number of them can cope with all of these at once. Similarly, in modern large-scale warfare there are too many encounters, planes, locations, events, supplies, etc., for the mind to keep track of. Conspicuous in these situations are a large number of changing interrelated events, variables, or conditions—just too much for a human mind to grasp at one time. For the computer, no sweat.

Summary   In general, when an engineer calls for a computer as part of his solution to a problem, it is for at least one of the four reasons just discussed. In some instances

Figure 7   Once a second the computer receives information from radar on the position of each plane in its traffic control region. It stores this information in the form of X-Y-Z coordinates. What you see here is the computer's "file" on Flight A715 for two successive seconds. The computer is doing likewise for other planes under its control at the time, switching its attention rapidly from one plane to the next. The computer can do more than simply store positions. For example, it can predict where each plane will be in the near future, as shown in the predicted position columns. It does this by using the present and immediate past positions of the plane to compute its speed and heading. Then on these bases it computes where the plane will be 5 seconds and 10 seconds later. In this way the computer can foresee potentially hazardous situations and alert traffic control personnel. Traffic controllers can request the computer to display this information graphically on their scopes, including the predicted positions, for specific planes or all planes in their area.

	Flight A715					Flight A715					
	$T-2$	$T-1$	T	$T+5$	$T+10$	$T-2$	$T-1$	T	$T+5$	$T+10$	$T-2$
x	9.12	9.03	8.94	8.49	8.04	9.03	8.94	8.85	8.40	7.95	8.94
y	27.04	27.03	27.02	26.97	26.92	27.03	27.02	27.01	26.96	26.91	27.02
z	2.14	2.14	2.14	2.14	2.14	2.14	2.14	2.13	2.08	2.03	2.14
	Positions 1 and 2 seconds ago		Position now	Predicted positions							
		(a)	$T = 9{:}22{:}02$				(b)	$T = 9{:}22{:}03$			

there is no choice; only the computer can handle the job. But in most applications it is a matter of economics—the engineer has compared costs and found the computer to be cheapest in the long run. The designer of the utility company's billing system compared the total costs of preparing bills by desk calculators, accounting machines, and computer and found the last method to be the least costly.

The major classifications of computer uses (pages 95 to 102) are neither all inclusive nor mutually exclusive. You may find a computer application that you cannot place in any of these categories, nor is there any point in laboring to do so. The prime purpose of this classification scheme is to provide a practical aid to you in spotting potentially profitable computer applications. Incidentally, computers are used in most fields of intellectual endeavor, including medicine, business, government, education, and the sciences, for these same basic reasons. Therefore familiarity with this chapter provides you with an appreciation of the pattern of computer use in *any* field.

The General Effects of Computers on Engineering

The computer is profoundly affecting engineering. For one thing it has resulted in expanded use of mathematics. An old timer in engineering may tell you, "Don't worry about all the mathematics you're getting in engineering school; most of it you won't need on the job." Beware; things have changed. Before computers many sophisticated mathematical techniques served mainly as vehicles for mathematicians to demonstrate their prowess in classroom and textbook. Practical use of these techniques was severely restricted because of the many man-hours required to solve the equations. Now this obstacle has been removed. A whole range of powerful mathematical techniques can be *applied* since the computer will do the dogwork quickly and at a tolerable cost.

The computer is cutting the routine, repetitive, tedious work of the engineer to a minimum. Take a quick look at pages 90–94 and note that in all cases the computer dramatically reduces the engineering-hours required. This is a welcome change indeed. *Not* that the engineer no longer must do pencil and paper calculations or draw or search through files; he still must, but not for prolonged periods. The computer is doing for engineering what the bulldozer has done for construction labor.

The labor-saving ability of the computer has another benefit. Before computers engineers frequently were forced to make gross, undesirable simplifications in many of their mathematical and simulation models. There is a very practical reason for this simplification: to yield equations that can be solved and simulation models that can be manipulated "by hand" in a reasonable period of time. This *is* a practical matter when you are under pressure to solve a problem as soon as possible, which is typical. Assume that you must predict the stresses in a structure you are designing. You have a fancy mathematical model for doing so, but it requires several days of computing if solved by hand. You also have a simplified model that will require ½ hour of hand computation. You know which model you would be inclined to use, even though the simpler model is less accurate. Of course you would compensate for the uncertainty in its stress predictions by using relatively large safety factors in your design. But now, with the computer, you can use the more sophisticated model and benefit from its more accurate predictions and the correspondingly slimmer factors of safety.

The computer is helping engineers make better use of what is known as illustrated on page 90, although we have only scratched the surface in this respect.

Finally, the computer has greatly extended the engineer's "ability to accomplish." Many of man's remarkable achievements in space travel, nuclear power, air transportation, and communications would be impossible or long delayed without computers.

The Future

The contemplated effects of time-sharing on engineering offer a hint of the impact computers are going to have. Visualize an engineering department with several hundred engineers distributed through many offices and buildings. They are served by a time-sharing computer system with forty input-output terminals, located so that every engineer has convenient access to the computer. All stations include a teletypewriter, some have graphic equipment similar to that appearing in Figure 2, some have tape and card readers, some have plotters and printers. From his terminal an engineer can initiate literature searches, call for information from the departmental files, solve equations, and do almost anything that is to be performed by computer. He can write a computer program, try it, "debug" it right then, and when he has a workable program instruct the computer to store it. Anyone wanting to use this program can call for it from any terminal and have it available instantly.

This system can have terminals thousands of miles away, in branch offices or other divisions of the company, tied into the one large computer by telephone lines. Such company-wide sharing of one computer gives everyone the use of a larger, more powerful machine, as opposed to buying a number of smaller machines and distributing them around the company. It also enables all engineers throughout the company to share technical information and computer programs, conveniently and quickly. Such systems are possible but far from commonplace.

Of course, much of what engineers do with computers in the future depends on time sharing. It would hardly be feasible for an engineer to sit all morning at a graphic terminal, sketching alternative designs, calling for different views of his drawings, making computations, etc., if his activity required the complete attention of a million-dollar computer. But it doesn't; in the four hours the engineer worked at the terminal he actually used only twelve minutes of computer time, in the form of many spurts of fractions of a second. During the remainder of the four hours the machine was available to the other 39 terminals.

The computer revolution in engineering is just beginning. There will be many new uses, many extensions of present ones. Incidentally, none of the computer applications described here is old hat; all are in the evolutionary stage. So there are plenty of opportunities and challenges for you.

It is useful to view the engineer and the computer he uses as a partnership in which man and machine complement one another, performing the functions for which each is best suited. Man excels at invention, at reasoning, at pattern recognition, at benefiting from experience. He adapts quickly to a remarkable variety of tasks. He excels at relatively short tasks because his "setup" time (e.g., time to get pencil and paper in hand) is usually short.

In contrast, the computer performs repetitive, routine tasks reliably and precisely, without boredom or fatigue, in about a millionth of the time required by human beings. It has to be instructed only once, and thereafter it follows those instructions any number of times without deviation. It has a perfect memory for endless details—and a memory uncluttered by useless information, for, when told to forget, it does so instantly and completely.

As improvements are made in computers and their programs, the machines will relieve engineers of more and more repetitive and routine tasks, allowing them additional time for creative and analytical thought. Thus the boundary between what human beings do better and what computers do better is gradually shifting, to the engineer's benefit.

Exercises

1 *Prepare a paper on one of these subjects, describing what the computer can do for engineers and what opportunities and challenges exist in the area you select.*
   (a) *Computer graphics.*
   (b) *Numerical control.*
   (c) *Information storage and retrieval by computer.*
   (d) *Digital simulation by computer.*
   (e) *Time sharing.*

8

THE DESIGN PROCESS:
PROBLEM FORMULATION

HERE is the general procedure for solving an engineering problem:

- *Problem formulation*—the problem at hand is *defined* in a broad, *detail-free* manner.
- *Problem analysis*—now it is *defined* in *detail*.
- *The search*—alternative solutions are accumulated through inquiry, invention, research, etc.
- *Decision*—the alternatives are evaluated, compared, and screened until the best solution evolves.
- *Specification*—the chosen solution is documented in detail.

This five-phase procedure, the *design process,* is described in this and the next four chapters.

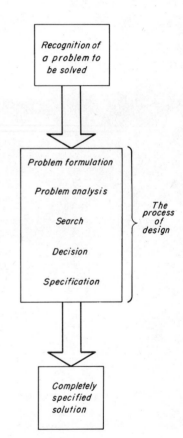

The process of design embraces the activities and events that transpire between the recognition of a problem and the specification of a functional, economical, and otherwise satisfactory solution to that problem. Design is the general process by which the engineer applies his knowledge, skills, and point of view in the creation of devices, structures, and processes. It is, therefore, the central activity in the practice of engineering. Whatever the engineer may be creating—a nuclear power generator, undersea craft, weapons system, dam, printing press, food-processing plant, or mechanical heart—he does so by means of the same basic design process. Now for a detailed verbal model of this procedure.

Problem Formulation

Do you advocate trying to solve a problem without knowing what the problem is? Surely not; yet this is exactly what you are prone to do, which is hardly conducive to effective problem solving. Certainly it makes sense to know what problem you are solving and whether it is worth solving, before rushing headlong into the details. It also makes sense to take a broad view of the problem *at the outset,* because once you are immersed in details breadth of perspective is virtually impossible to achieve. Therefore, the primary objectives of problem formulation are to decide in general terms what the problem is, to determine whether it warrants attention, and to get a broad view of that problem when it is best and easiest to do so. Obviously these are matters that should be known at the outset. This crucial phase of the design process requires only a small proportion of the total time devoted to a problem, a fact that belies its importance.

Rarely is the true problem laid before the engineer. He must determine for himself what the problem is. This is often difficult because its nature is ordinarily obscured by much irrelevent information, by solutions currently in use, by misleading opinions, and by unprofitable customary ways of viewing the problem. This matter is not aided by the fact that in college problems are usually presented to students in an unrealistically pure form, so that fledgling engineers are unaccustomed to and unskilled in defining problems. In view of these situations and of the consequences of neglected or ineffective problem definition, it behooves you to begin *now* to develop your skill in formulating real-world problems.

A Case in Point   The management of a large company that distributes livestock feeds is concerned about the relatively high cost of handling and storing its products. An engineer has been assigned to the problem to seek a significant cost reduction. At present the materials are bagged and stored by the method diagrammed in Figure 1.

A common tendency is immediately to begin trying to think of possible improvements in the existing solution (there usually is one). In this problem it is tempting to begin by scrutinizing the solution described in Figure 1, seeking improvements that may make the process more economical. The person who does this will immediately become concerned with such matters as equipment for filling, weighing, and sewing bags, arrangements of facilities, ways of transporting the heavy sacks, means of combining operations, and other possible improvements.

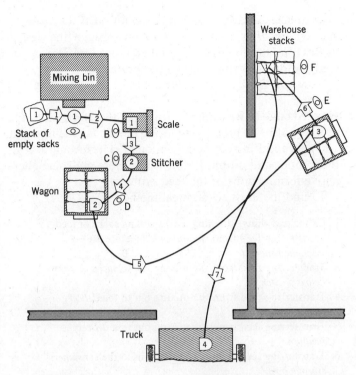

Warehouse
stacks

Mixing bin

Scale

Stack of
empty sacks

Stitcher

Wagon

Truck

1 Stacked sacks await filling.

2 Man C folds and stitches the top of the bag.

1 Man A lifts empty sack from stack
and places it under spout for filling.

4 Man D takes the bag and loads it on wagon.

1 Man A fills the 100–pound sack by
gravity feed, manually controlling
the rate of flow.

5 Loaded wagon is pushed to warehouse.

2 Man A hands the bag to man B.

6 Bags are stacked by men E and F.

1 Man B checks the weight and adds or
removes material when necessary to adjust
the weight to approximately 100 pounds.

1 Bags are stored awaiting sale.

3 Man B hands the bag to man C.

7 Bags are loaded on waiting truck,
two or three at a time by handtruck,
then delivered to consumer.

*Figure **1*** *The present method of filling, storing, and loading bags of feed.*

This is exactly what *not* to do in approaching a problem—immediately to become embroiled in the process of generating solutions (which has its place, later). Note that in doing so the person is generating solutions *to a problem that he has failed to define.* Believe me,—you'll pay if you make a habit of this.

Note: the current solution to a problem is not the problem itself. This statement seems obvious, yet you probably do this very thing: attack the present solution, not the problem. There is a subtle but crucial difference between picking at the current solution in an effort to eliminate inadequacies, and starting with a definition of the problem and synthesizing a superior solution through the design process. In the long run the latter procedure is a major contributor to superior design performance.

Now you know what not to do.

How Should It Be Done?   Right at the start, state the problem at hand broadly, ignoring details for the moment and concentrating on identification of states A and B (call them input and output if you wish). Here are some alternative formulations of the feed problem.

To find the most economical method of . . .

Do you see differences in the solutions that might result from these various formulations?

1. Filling, weighing, stitching, and stacking sacks of feed.
2. Transferring feed from the *mixing bin* (state A) to *stockpiled sacks in the warehouse* (state B).
3. Transferring feed from the mixing bin to *sacks on the delivery truck.*
4. Transferring feed from the mixing bin to the *delivery truck.*
5. Transferring feed from the mixing bin to a *delivery medium.*
6. Transferring feed from the mixing bin to the *consumers' storage bins.*
7. Transferring feed from the *storage bins* of the *feed ingredients* to the consumers' storage bins.
8. Transferring feed from the *producer* to the *consumer.*

Formulation 1 is unacceptable; it fails to identify states A and B and it includes restrictions, "filling, weighing, stitching, and stacking," which have no place in a problem formulation. Note that these restrictions are characteristics of the current solution. Formulations 2–8 are acceptable *but they are not equally advisable.* This array of formulations and the fact that the probable consequences of following each are quite different raises an important matter referred to as breadth of the problem formulation.

Breadth of Your Problem Formulation   In formulations 2 and 3 it is assumed that the feed is in sacks at state B. In formulation 4 only "truck" is specified, thus opening the problem to solutions not involving sacks. In formulation 5 only "delivery medium" is specified for state B, opening up additional possibilities not involving trucks. This trend toward a less specific definition of states A and B continues until only producer and consumer are specified, leaving the way open for a wide variety of methods of handling, modes of transportation, package types, etc. Thus it becomes apparent that as the specifications assumed for states A and B become more general, the alternative solutions available to the designer become more numerous and varied. Strive to keep your formulation as general as the importance of the problem justifies. Failure to follow this policy will cause whole realms of profitable possibilities to be unnecessarily excluded from consideration. Most persons trying to solve the feed problem would automatically and unjustifiably assume state B to be stockpiled sacks in the warehouse, and complete the whole design process without realizing that only they themselves have limited the problem to this extent.

In formulation 2, state B goes only as far as the warehouse stockpile (Figure 2). Formulation 3 extends state B to the truck, and formulation 6 extends it to the consumer. In formulations 7 and 8, state A is extended. In each of these instances the problem is being extended to embrace more of the total problem. In general, strive to formulate your problem *to include as much of the total problem as the economics of the situation and organizational boundaries will permit*. The more a total problem is split into subproblems to be solved separately, the less effective the total

Figure 2 Alternative formulations of the feed distribution problem, illustrating progressively broader formulations of a problem.

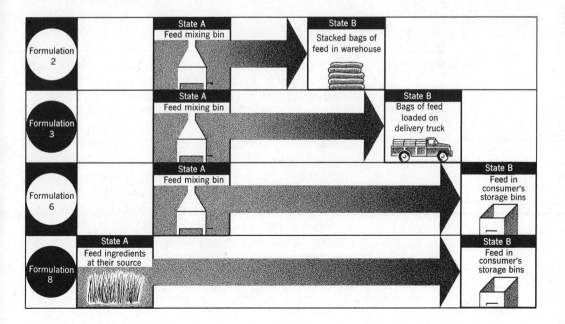

solution is likely to be. If bagging of feed is treated as one problem, transportation to and stacking in the warehouse as another, transportation to the consumer as another, and unloading of trucks as another, the overall feed distribution system that eventually evolves will probably be far from optimum. Treating this problem broadly is very likely to result in a superior over-all system.

The detail in which states A and B are specified and the proportion of the total problem that they encompass will henceforth be referred to as the breadth of the problem formulation. Formulation 8 of the livestock feed problem is certainly broad in contrast to formulation 2.

Importance of a Broad Formulation   The engineer assigned to this project succeeded in freeing himself from the limitation of using sacks, and thereby opened the problem to the possibility of handling feed in large bulk. Also, his formulation encompassed delivery to the consumer, which opened the way to delivering feed in bulk right into the farmer's storage bins. The result: after many years of back-breaking loading and unloading of heavy sacks, dealers now deliver feed by blowing it through a hose leading from a bulk-delivery truck, which is a large "feed bin on wheels," directly into the farmer's storage bins. (This system closely parallels the usual method of delivering household fuel oil.)

Broad treatment of problems that previously were attacked in piecemeal fashion can pay off handsomely. We are surrounded by problems that are unsatisfactorily solved mainly because the solvers are traditionally "narrow-sighted." This statement applies to education, business, medicine, and every other field, as well as to engineering. The reason I am making this plea for broad problem formulations is that the probability of vastly improved solutions is so high. There are great opportunities for engineers who will attack problems in an unconventionally broad manner. The following case study illustrates what I am advocating.

City X is plagued by a severe parking problem. Forty per cent of its commercial district is parking area. This has prompted city officials to hire a consulting engineer to design a 600-car, multistory parking facility. It will be one of a series of similar facilities to be constructed in the congested area.

(This city's dilemma arose primarily because several years earlier problem solvers with an insufficiently broad view of the community's problem had called for the construction of a network of superhighways, making it quick and simple to travel by automobile between the suburbs and the downtown area. The result was predictable. Commuters took advantage of this high-speed funnel to the heart of the city and flooded the commercial district with automobiles.)

Before specifying the details of the desired facilities, the engineer devotes some thought to the problem to which the proposed structures are presumably the solution. (Note that the engineer has been given the city fathers' solution to the problem. His task is to detail this solution so that it is structurally, economically, and functionally sound.) The engineer views the underlying problem as that of transferring a large segment of the population between its place of residence and its place of business. There is a major difference between this formulation of the *problem* and the restricted *solution* given to the consultant. His broad formulation opens up the problem to a whole realm of promising solutions. One is a high-speed transit system. Of course, nothing in this engineer's formulation precludes the possibility of a different type of urban community that reduces the need for mass transportation.

The engineer does the ethical thing. He informs the city officials that in his opinion increased parking capacity is not the obvious answer to the problem and that he is reluctant to design the proposed facilities. Instead, he outlines his view of the problem and some alternative solutions stemming from it. Adopting a broad view of the problem and standing by it when this course is in the best interest of the client is a mark of a professional engineer.

How Broadly Can You Formulate a Problem?   This is *your* decision to make. A problem formulation is a point of view—the manner in which you perceive the problem. It may be no more than some thoughts or some scribbled notes. It is not irrevocable; it can be changed if you find this necessary or desirable. Therefore you should formulate problems broadly; it is your prerogative—in fact your professional obligation—to do so. You are selling yourself and your employer short if you don't. However, a broad formulation in your mind is one thing; to what extent you are able to apply it through the remainder of the design process is quite a different matter. Pursuance of a broad formulation can bring you into direct conflict with decisions already made by your client or employer, or it can lead you into decision areas that are considered the responsibility of other persons in the organization. The engineer assigned to the feed distribution problem encountered some resistance when he attempted to pursue his broad "producer-to-consumer" formulation. He had to persuade persons responsible for such decisions to give up the idea of individual sacks, to change the methods of warehousing, to alter sales policy, and the like. He succeeded, but for a variety of possible reasons someone might well have told him to mind his own business, thus forcing him to pursue a narrower formulation that was not in the best interest of the enterprise.

To what extent you are justified and able to *pursue* a broad formulation depends on the scope of your responsibilities, the importance of the problem, and the limit (if any) on time and money that can be devoted to the problem.

Methods of Formulating a Problem   A problem can be satisfactorily formulated verbally or diagrammatically, on paper or in your mind. In many instances a few words will suffice (page 110). Or maybe you prefer a simple diagram (page 111). The black-box method of viewing a problem is a diagrammatic formulation. The usefulness of this approach can be illustrated by applying it to a type of problem that is often unsatisfactorily defined: an information-processing problem. An office that handles reservations for seats, for instance, on a plane or in a theater, is an information-processing system. The potential customer comes to the ticket agent with a request in terms of number of seats, a date, and perhaps other specifications. These constitute the input to the black box. The output is also information, in the form of confirmation of the request or else a quotation of the alternatives that are available. In the problem-formulation stage what transpires within the black box is not known or of interest; it replaces the details that we are trying to avoid at this stage, and herein is the key to its helpfulness. This is a valuable way to view an information-processing problem or *any* type of problem. The simplicity

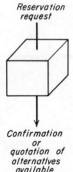

Reservation request

Confirmation or quotation of alternatives available

of the black-box approach belies its effectiveness as a problem-solving aid.

It must be occurring to you about now that I am not giving you hard and fast rules for problem formulation. None are justified. There is no such thing as *the* correct formulation of a given problem, but there certainly are more and less profitable ones. The best I can do is to offer guidelines—which I have already done—and cite examples (Figure 3 should help). It is up to you to benefit from these and from your own experience in developing your problem-formulating skill.

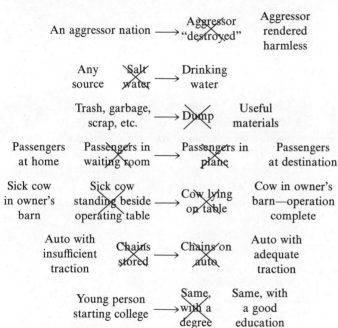

Figure 3   Some formulations of familiar problems, showing a broadening of each. Try to visualize for each example the possible benefits of the broader formulation. (In the first instance the broader view opens the problem to nondestructive means of subduing aggressors, such as methods of temporarily incapacitating them or their weapons.)

Summary   It has been said that a problem properly defined is virtually solved. Although this claim is exaggerated, it does serve to dramatize the crucial nature of this phase of the design process. A problem can be formulated in varying degrees of breadth. These may range from a very broad definition, which maximizes the number and scope of alternatives that can be considered, to a formulation that offers very little latitude in the way of possible solutions. You have your choice.

The input to the problem formulation phase is information, vague and cluttered with irrelevant and misleading facts, on what is needed or wanted. The output, a profitable formulation of the problem, becomes an input to the next phase of the design process, problem analysis.

Exercises

1   *Identify states A and B for the prime problem faced by each of the following individuals. Make assumptions where necessary.*
   (a)   *Newspaper delivery boy.*
   (b)   *Mountain climber.*
   (c)   *Cook.*
   (d)   *Firefighter.*
   (e)   *Teacher.*
   (f)   *Television repairman.*
   (g)   *Pottery maker.*
2   *Formulate the problem to which each of the following is a solution. (For example, the usual input to a gearbox is a shaft rotating at one speed and the output is a shaft rotating at a different speed.)*
   (a)   *Public address system.*
   (b)   *Electric iron.*
   (c)   *Air conditioner.*
   (d)   *Telephone system.*
   (e)   *Interstate oil pipeline.*
   (f)   *Bottling plant.*
   (g)   *Airport.*
   (h)   *Oil refinery.*
   (i)   *Snowplow.*
   (j)   *Commuter train.*

9

THE DESIGN PROCESS: PROBLEM ANALYSIS

AN appliance manufacturer has tentatively decided to market a new type of clothes washer. This machine is to perform the usual tasks expected of it and also is to serve as a home dry-cleaning unit. In addition, the management has decided that:

1. This unit *must* not be larger than 30 inches wide, 38 inches high, or 30 inches deep.
2. It *must* operate on 60-cycle, 115-volt alternating current.
3. It *must* be approved by Underwriters Laboratories.
4. The cost of manufacture *must* not exceed $125.
5. It *must* satisfactorily process all natural and synthetic textile materials.
6. It *must* be foolproof against blunders in operation.

The engineer assigned to design this multipurpose machine based his analysis of the problem on considerable deliberation, investigation, and consultation, especially with company executives and marketing experts who are in close touch with consumer preferences. His analysis, shown in Figure 1, is discussed in the following sections.

Specification of States A and B In formulating this problem it is sufficient to identify state A simply as soiled fabrics and state B as the same fabrics clean. However, to solve the problem it is necessary to learn more about the input and output. Thus during this stage of the design process detailed qualitative and quantitative characteristics of states A and B are determined, as demonstrated in Figure 1.

Very few characteristics of states A and B are constants. The amount of clothes a person places in a washer varies from load to load, so do the types of fabrics, the amount

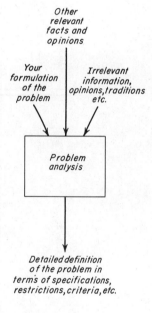

Other relevant facts and opinions

Your formulation of the problem

Irrelevant information, opinions, traditions etc.

Problem analysis

Detailed definition of the problem in terms of specifications, restrictions, criteria, etc.

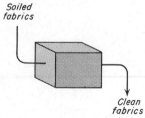

Soiled fabrics

Clean fabrics

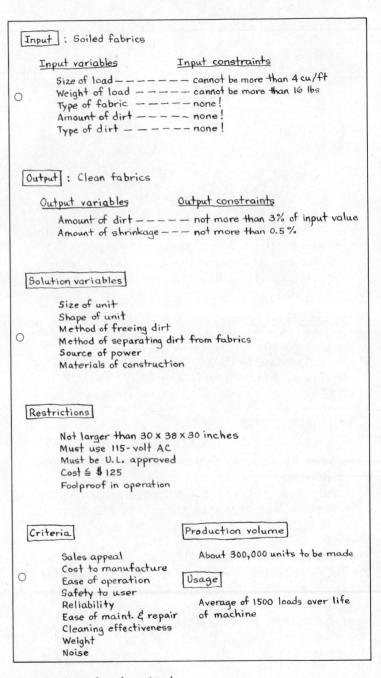

Input : Soiled fabrics

   Input variables   Input constraints

      Size of load — — — — — — cannot be more than 4 cu/ft
      Weight of load — — — — — cannot be more than 16 lbs
      Type of fabric — — — — — none!
      Amount of dirt — — — — — none!
      Type of dirt — — — — — — none!

Output : Clean fabrics

   Output variables   Output constraints

      Amount of dirt — — — — — not more than 3% of input value
      Amount of shrinkage — — — not more than 0.5%

Solution variables

      Size of unit
      Shape of unit
      Method of freeing dirt
      Method of separating dirt from fabrics
      Source of power
      Materials of construction

Restrictions

      Not larger than 30 x 38 x 30 inches
      Must use 115-volt AC
      Must be U.L. approved
      Cost $\leq$ \$125
      Foolproof in operation

Criteria   **Production volume**

      Sales appeal   About 300,000 units to be made
      Cost to manufacture
      Ease of operation   **Usage**
      Safety to user
      Reliability   Average of 1500 loads over life
      Ease of maint. & repair   of machine
      Cleaning effectiveness
      Weight
      Noise

Figure 1 A page from the engineer's notebook, showing his analysis of the cleaning-machine problem.

and type of dirt, and so on. (Examples from other problems: the ore input to a steel mill varies in chemical make-up; the electrical output of a power plant certainly varies over a period of time.) These dynamic characteristics of states A and B are called *input variables* and *output variables.*

Often there are limits on the degree to which input and output variables can fluctuate—for example, "the input variable weight of load cannot exceed 16 pounds," which can be expressed symbolically as $O < I_w \leqq 16$ pounds. This is called an *input constraint,* and the equivalent for state B is an *output constraint.* Before an engineer can satisfactorily solve a problem he must have reliable estimates of input and output variables and constraints.

Restrictions   A restriction is a solution characteristic previously fixed by decision, nature, law, or any other source the problem solver must honor. Thus the decisions quoted on page 117 appear in Figure 1 as restrictions. Each limits the choices open to the problem solver. Some restrictions limit his choice to a range of values: "the machine cannot be greater than 30 by 38 by 30 inches"; others *fix* a solution characteristic: "it must operate on 60-cycle, 115-volt alternating current." Thus solutions larger than 30 by 38 by 30 inches are ineligible; any other type of power source is ruled out. Usually such decisions are made by the engineer's employer.

Not all restrictions are accepted by the engineer. For instance, it is possible to design a cleaning machine that will satisfactorily process all types of fabrics, but the expense involved in developing and manufacturing a truly general-purpose machine will be high. The engineer believes the probable selling price of a machine that would process all fabrics is disproportionately higher than the probable price of a machine that processes *almost all* fabrics. He is convinced that the "process-all-fabrics" decision is economically unsound. Under these circumstances the engineer must decide whether he should accept this restriction or seek to have management reconsider and possibly revoke its original decision.

Furthermore, it is apparent to the engineer that two of the imposed restrictions are incompatible: a general-purpose machine at a manufacturing cost of $125 or less.

One or both of these restrictions must be relaxed if a solution is ever to be found.

Thus you probably will not adhere to all restrictions imposed, usually with the agreement of those who imposed them, because some cannot be heeded and others can be satisfied only at a disproportionately high price. You are naive if you assume that all restrictions represent optimum decisions to be accepted without challenge. *Most* of the decisions made by executives, engineers, and others are suboptimum to *some* degree. This results from the element of chance inherent in the search for alternative courses of action, the relatively short span of time available for making decisions, the predominant role of judgment in making real-life decisions, the many future implications and consequences that are unforeseen, the degree to which problems are often subdivided and attacked as relatively independent subproblems, and the fact that few decisions are made on a completely objective basis.

Therefore do not *automatically* accept all restrictions given. Many a profitable innovation owes its existence to an engineer who did not blindly accept every restriction as sound and irrevocable.

Fictitious Restrictions   Take the problem in which you are to connect these nine dots by not more than four straight lines without removing your pencil from the paper while drawing the lines. Some people cannot solve this problem and others require a long time to do so, because they unjustifiably and probably unawaringly rule out the possibility of extending the lines beyond the square formed by the dots. They behave as if this were not permitted, even though no such restriction was mentioned in the statement of the problem. This unjustified, undesirable ruling out of a perfectly legitimate alternative or group of alternatives is a fictitious restriction.

Most fictitious restrictions are not explicit decisions to disregard certain possibilities. Instead the problem solver automatically acts as if certain alternatives have been ruled out. Many would proceed to solve the feed problem (page 109) as if the feed must be handled in sacks, even though no one said this must be done. The fact that fictitious restrictions are not ordinarily arrived at through conscious deliberation is the main source of their elusiveness. If they are explicitly stated, their imaginary and often absurd nature becomes quite obvious. Referring again to the feed problem, suppose I state some of the characteristics of the current solution as restrictions, which they are not: "feed *must* be placed in sacks; the sacks *must* be handled individually." These are nothing more than characteristics of the present system with the words "must be" substituted for "is" and "are." The present solution is a common source of self-imposed, imagined boundaries; the tendency to accept what is as what must be is strong. Because everyone has this tendency and because the elimination of a fictitious restriction usually opens up a problem to worthwhile solutions, you should make a special effort to watch out for this pitfall.

Solution Variables Alternative solutions to a problem differ in many respects. Solutions to the cleaning-machine problem differ in such features as size, shape, method of freeing dirt from fabrics, type of mechanism, materials from which the machine is constructed. The ways in which solutions to a problem can differ are called *solution variables*. The final solution to a problem consists of a specified value for each of those variables; a certain size, a certain shape, and so on.

Please be sure you understand the purpose of determining restrictions and solution variables. The purpose is not to learn all the ways you are restricted; *it is to learn in what ways you are not restricted, and subsequently to take advantage of this freedom in your search for solutions.* To aid you in this cause, I recommend that you first identify all solution variables; then determine which ones are justifiably fixed or limited.

Criteria The criteria that will be used in selecting the best design should be identified during problem analysis. Actually, criteria change very little from problem to problem; construction cost, safety, reliability, ease of maintenance, and the like apply in almost every case. But what does change significantly are the relative weights of these criteria. Hence in most cases the engineer's main task with respect to criteria is to learn the relative importance attached to various ones by officials, customers, clients, citizens, or others concerned. This information is important; the following example will illustrate why. Assume that safety is to be a heavily weighted criterion in the design of a new-model rotary lawn mower. Knowing this, the designer will consider different materials, mechanisms, cutter types, discharge methods, etc., than those which he would otherwise investigate. A heavily weighted criterion affects the types of solutions emphasized in the search for alternatives, and this fact should be known before that search begins.

Usage   If a river is to be crossed only rarely at a given spot, a bridge is obviously not the solution that minimizes total cost (design plus construction plus crossing costs). On the other hand, if millions of persons need to cross the river at this spot over a period of time, a rowboat is not the preferred method with respect to the total cost criterion. The number of times that the transformation associated with a problem is to be repeated becomes significant whenever *total* cost (the cost of arriving at a solution plus the cost of physically creating it plus the cost associated with using it) is a matter of concern. And when is this not the case? Recall the reed switch problem (page 6). You can imagine what type of manufacturing methods would be used if only several hundreds (instead of many millions) were needed.

Before an engineer can solve a problem intelligently, he must determine the expected *usage*—the extent to which the solution is to be employed—since this strongly affects the type of solution that will be optimum. For the cleaning-machine problem the usage is 1500 average-size loads over the life of the appliance.

Production Volume   Suppose that only ten cleaning machines are to be built. Under these circumstances the designer would care little about the manufacturability of his creation. If nonstandard expensive components were specified and hand methods of fabrication were required, these would be of little concern as long as the volume were only ten machines. However, 300,000 is a different story; under these circumstances the engineer will be vitally concerned with the manner in which alternative designs affect manufacturing cost. This number, referred to as *production volume,* has a significant effect on the type of solution that is optimum for the problem, and obviously should be known before you start looking for solutions.

Problem Analysis—the General Form Problem analysis involves much information gathering and processing. The result is a problem definition—in detail, as illustrated by Figures 1 and 2—which hopefully maximizes your chances of finding the optimum solution.* Now you are ready to begin a *search* for that solution.

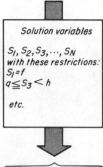

$Input$, with input variables
$I_1, I_2, I_3, ..., I_N$
and these constraints:
$I_1 = a$
$0 < I_2 < b$
$c \leq I_4 \leq d$
etc.

Solution variables

$S_1, S_2, S_3, ..., S_N$
with these restrictions:
$S_1 = f$
$q \leq S_3 < h$

etc.

$Output$, with output variables
$O_1, O_2, O_3, ..., O_N$
and these constraints:
$O_2 > j$
$O_5 < k$
etc.

Criterion = C
Usage = U
Production volume = V

Exercises for Chapter 9

1 *Assume that you are designing the following:*
 (a) *Vegetable cannery.*
 (b) *Structure for crossing the English Channel.*
 (c) *Transoceanic telephone cable.*
 (d) *Power-generating station using coal as fuel.*
 How do you formulate each of these problems? What information concerning input and output would you gather during your analysis of each problem? For each, list some major criteria that you believe should be employed. Identify the major solution variables for each.
2 *What restrictions can you isolate in each of the case studies presented in Chapter 2?*

Figure 2 Summary of the types of information which you should gather in your problem analysis. Lower-case letters are substituted for what would ordinarily be actual numbers. Given this information, it remains for you to find the combination of values for S_1, S_2, S_3, ... S_N that maximizes C and satisfies all constraints and restrictions.

Exercises for Chapters 9 and 10

(The following problems involve creative thinking as well as problem definition and therefore are appropriate for this and the next chapter.)

1 *One portion of a large oil company warehouse is devoted to the storage, packing, and shipping of road maps to service stations. The current procedure is diagrammed in Figure 3. The stored maps are removed from their cases (200 to a case, 8 bundles of 25 maps each), and a moderate supply for each state is stacked on the open shelves. The packer, with order slip in hand, fills the customer's order by picking the requested type and quantity of maps*

* The problem nomenclature summarized in Figure 2 is adequate for your initial design experience. Eventually, however, you may want to be more rigorous in your problem analysis. Therefore I have included some refinements and extensions in Appendix C.

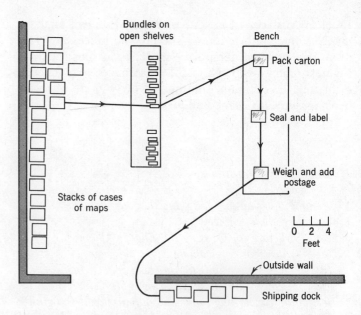

Figure 3   The present procedure for assembling and packing map orders.

Table 1

Types of map requested in the order	Per cent of orders received
1	3
2	3
3	5
4	8
5	16
6	19
7	16
8	14
9	8
10	4
11	3
12	1

from the shelves and assembling them in a carton on the bench. When the order is assembled, he slides the carton to the sealing and labeling station and performs these operations. Then he slides the carton to the next position in order to weigh the shipment and add the required postage. He then carries the completed order to the shipping dock. Under this method, the typical order requires an average of 10 minutes for completion. The wage rate is $1.94 per hour. There are approximately 13,000 orders per year.

Twelve different maps must be stocked. An analysis of the orders received over a period of time gave the results shown in Table 1. A service station cannot requisition more than a total of 500 maps in any one order. A quantitative analysis of orders received over a period of time produced the results shown in Table 2.

As the engineer assigned to improve this procedure, what do you recommend? (The process must be performed in the same part of the warehouse that is presently used.) Adequately describe the workroom layout, the procedure, and the equipment that you propose. Also provide an outline of the manner in which you arrived at your solution, including your problem formulation and analysis and the alternatives you considered.

2   *As a consulting engineer, you have been engaged by a large city to provide the general specifications for a new, high-speed, high-capacity transit system to interconnect the suburban and downtown areas. Sketches of the general features of your proposal will be satisfactory.*

3   *At a clinic for large animals a number of operations must be performed on horses, cows, and bulls. As you can well imagine, getting one of these animals to lie on an operating table is a real challenge. Specify the general features of a device or system that you recommend for placing these animals on their sides on the operating table.*

Table 2

Number of maps requested in the order	Per cent of orders received
0–50	7
51–100	11
101–150	18
151–200	26
201–250	13
251–300	10
301–350	6
351–400	3
401–450	4
451–500	2

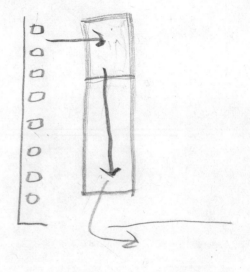

10

THE DESIGN PROCESS: THE SEARCH FOR ALTERNATIVE SOLUTIONS

SOLUTIONS will probably occur to you as you define a problem, but as byproducts, not as the objective of your efforts. In this phase of the design process you actively seek alternative solutions. You become engaged in what is truly a *search,* of the mind, of the literature, and of the world about you. Man's vast accumulation of knowledge provides ready-made solutions for some parts of most problems. Searching for these is a relatively straightforward process of exploring your memory, books, technical reports, and existing practices. There is a second major source of solutions—your own ideas, the fruits of the mental process called *invention.* You will rely heavily on your ingenuity to solve the many aspects of problems not covered by existing technical and scientific know-how. Unfortunately, however, inventing solutions is not as straightforward and controllable as looking up ready-made ones; this you recognize from your own problem-solving experience—ideas do not ordinarily flow forth whenever you want them. Consequently it pays to devote special attention to improving your inventive ability.

Inventiveness refers to the number of worthwhile solutions a person is capable of conceiving. *Your inventiveness depends on* your attitude, your knowledge, the effort you put forth, the method you employ in seeking out ideas, and your aptitude (inherited qualities that affect your inventiveness). *Note that you control four of these five determinants; therefore it is within your power to improve your inventive ability.* You *can* over a period of time improve your attitude and increase your knowledge. You *can* in-

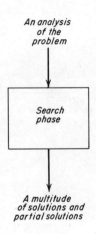

An analysis
of the
problem

Search
phase

A multitude
of solutions and
partial solutions

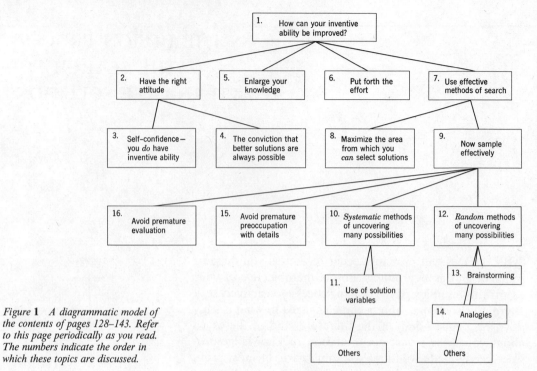

Figure 1 *A diagrammatic model of the contents of pages 128–143. Refer to this page periodically as you read. The numbers indicate the order in which these topics are discussed.*

crease your effort. You *can* substantially improve your method of searching for solutions. This should be good news for those of you who are not natural creative geniuses; for, although aptitude is not under your control, this can ordinarily be compensated for by the remaining four determinants, which you *can* influence. How this can be achieved is explained on succeeding pages, as summarized by Figure 1.

Improving Your Inventiveness—the Right Attitude First and foremost, you must believe that you have the potential to be creative. Why are some persons noticeably more inventive than others? You are probably thinking, "They were born that way." But who is to say what the relative contributions of the five determinants are to a particular person's creative performance? It is impossible to tell. An extraordinarily creative person could just as well be that way not because he was exceptionally endowed at birth but because his attitude, knowledge, and methods are usually good and because he works hard. It may be that a few rare persons are exceptionally endowed through heredity, but it is also *true that everyone inherits some inventive ability and that very few people fully exploit what they have.* It is very likely that insofar as this trait is concerned you were not shortchanged at birth. So do not be concerned about whether you have any special aptitude for invention; concern yourself rather with putting what you have to maximum use. Look at it this way: if you assume that you have no creative aptitude, you will probably live up (actually *down*) to your expectation. If your attitude is positive—you *know* you have potential—then all you must do is use it!

Another aspect of this matter of attitude is an insatiable drive to find better solutions. Rarely if ever will you work on an engineering project in which you are able to isolate all solutions. Usually you will run out of time before you run out of possibilities. The consulting engineer who developed a baggage-handling system for an air terminal isolated a variety of methods of transporting passengers' luggage, numerous sorting systems, and many pick-up arrangements. He finally ended his search because other problems were awaiting his attention, not because he had exhausted all possibilities. Nor did he believe at any time during this project that he had thought of all solutions. A creative engineer, regardless of how many solutions he has conceived, quite justifiably *assumes* that more and better ones remain to be discovered. He not only assumes they exist; he goes after them as long as time permits.

The engineer is "alternatives hungry"

The first step, then, in becoming more inventive is to develop a positive attitude toward the possibility of *your* finding better solutions to any problem you face. Each time you isolate an additional solution, set out again to find a better one, and continue until you are forced to give up because of a project deadline or the pressure of other problems awaiting your attention.

Improving Your Inventiveness—Expanding Your Knowledge When you get an idea, you have combined two or more bits of knowledge in a manner that is novel for you. It is a reorganization of your knowledge; an idea does not come from a void. Therefore the larger your store of knowledge, the more raw material you will have available for generating solutions. Furthermore, the broader the range of subjects this knowledge embraces, the better are the prospects for unique and often particularly effective ideas (one of several arguments for broadening your knowledge).

Ducted propeller

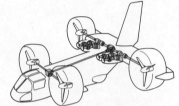

Figure 2 Speaking of alternative solutions! In addition to the familiar helicopter and the plane described on page 24, the VTOL designs on this and the following pages are also being considered for military and commercial use. The story is the same for almost every engineering problem: possibilities and more possibilities, it's difficult to run out of them. This study in alternatives is intended to arouse your "alternatives consciousness." (Incidentally, there are still other VTOL designs under study.)

(Courtesy of Textron's Bell Aerosystems Company.)

130

Tilt wing

(Courtesy of Ling-Temco-Vought, Inc.)

(Courtesy of Hughes Tool Company.)

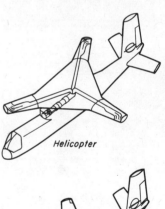

Helicopter

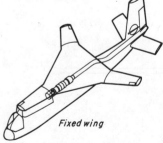

Fixed wing

Rotating wing

Deflected jet

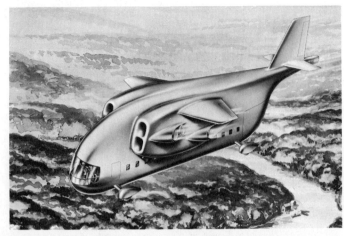

Deflected jet
(the "flying pig")

(Courtesy of Hawker Siddeley Aviation Limited.)

Stowed rotor

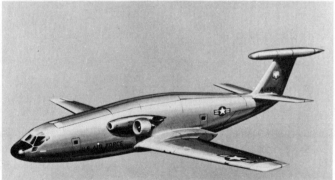

*(Courtesy of Sikorsky Aircraft Division,
    United Aircraft Corporation)*

133

Tilting rotor

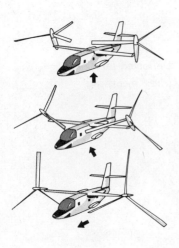

(Courtesy of Bell Helicopter Company.)

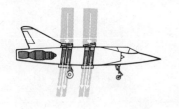

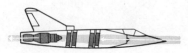

Vertical fuselage engines

(Courtesy of Avions Marcel Dassault.)

(Courtesy of Ling-Temco-Vought, Inc.)

Propulsive wing

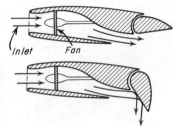

(Courtesy of Dornier GMBH)

Wing-tip engine pods

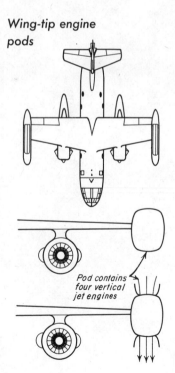

Pod contains
four vertical
jet engines

School is not the only source of such knowledge; observation, conversation, reading, and other forms of lifelong learning are also very important.

Improving Your Inventiveness—Putting Forth More Effort
Occasionally a creative idea strikes you when you apparently aren't thinking or even caring about the particular problem involved. But there is a big difference between an occasional flash of genius and the consistent production, under pressure, of ideas and more ideas to solve a given problem in the time available. This type of productivity takes effort. You won't find many really creative persons who are not hard workers. Hence to maximize your inventiveness you must be willing to exert effort.

Improving Your Inventiveness—Employing Effective Methods of Search   It will pay you to devote careful attention to your method of searching for solutions. You will understand why after you realize what you are prone to do in problem solving *if* you don't attend to this matter. The following analogy will highlight some of the common pitfalls, difficulties, and flaws in procedure.

Visualize the X's in Figure 3 as points in space, each representing a solution to a problem at hand. Assume that widely separated points represent radically different solutions. Hopefully, when trying to think of solutions to a problem, you would start somewhere in this space and move to progressively better solutions until a time limit or perfection ended the search. Limitations of the mind prohibit such effective performance. In fact, this search process is typically hit or miss, suffering from objectionable degrees of regression, inefficiency, and lack of direction.

All too often you will start with the present solution, the point *S* in Figure 3, and proceed from one point (idea) to another in a manner indicated by the arrowed path. Notice that the jumps tend to be relatively small, so that ideas cluster undesirably about the current solution. Why do your alternative solutions tend to resemble the present way(s) of solving the problem?

One reason could be that you fail to put forth sufficient effort. Another possibility is that you are in effect looking for nothing more than modifications of the current solution, rather than for a variety of basically different ones.

It is quite possible too that even if you are trying hard,

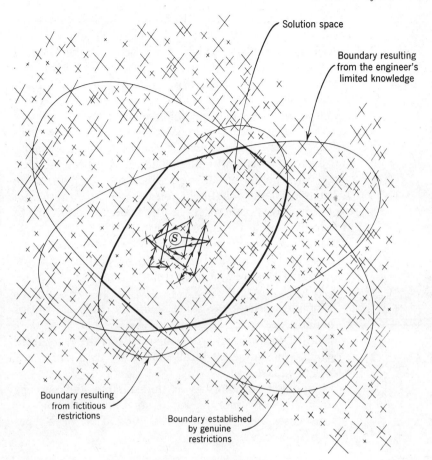

Solution space

Boundary resulting
from the engineer's
limited knowledge

S

Boundary resulting
from fictitious
restrictions

Boundary established
by genuine
restrictions

your thinking is in a rut because the customary solution to a problem has a natural power of attraction, especially so if this solution has a long history of use or if you are intimately familiar with it. After you have "lived" with a certain solution for a while it becomes a formidable block to original thought. Such familiarity leads to narrowness and inflexibility of thought and could easily cause the clustering shown in Figure 3. The world abounds with illustrations of this. A good example is found in the history of man's attempts to fly. The solution with which men were familiar was the wing action of birds and insects. Almost inevitably, they tried to fly by employing all sorts of unsuccessful and often disastrous wing-flapping schemes. In time they freed their thinking from "the stranglehold of the familiar."

Figure 3   The arrowed path represents the manner in which you are prone to search for a new solution to a problem —unless you are very unusual or have devoted careful attention to your search method. (The boundaries on this solution space are explained later.)

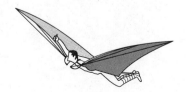

Another cause of idea clustering is the tendency to be conservative. There is security in sticking close to what has survived years of use. This fact, coupled with the common assumption that large investments in solutions are undesirable or forbidden, encourages you to come up with a solution similar to the prevailing one.

These and other detrimental tendencies will inhibit your ability to invent *unless* you take measures to minimize their effects. Here, then, are such measures, intended to maximize the number and value of alternative solutions that you can generate in solving a given problem. These are matters of method that help you to realize the most from your inventive potential. (I trust that you are continuing to relate the text to the diagram on page 128.)

First, maximize the number and variety of solutions from which you can sample. This you do by pushing back the boundaries (Figure 3) arising from:

- Genuine restrictions—some solutions really are out of bounds.
- Limited knowledge—your mental store of facts from which you synthesize ideas embraces only a fraction of all knowledge.
- Fictitious restrictions—you unjustifiably and perhaps unintentionally rule out some worthwhile solutions. In most cases this is the most restricting of the three types of boundaries.

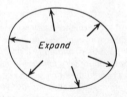

Push the limits back: eliminate the fictitious restrictions, screen the real ones, and supplement your knowledge relevant to the problem at hand.

Then take full advantage of this expanded solution space; search it effectively. To do so you should sample all areas of possibility that offer promise of containing the optimum solution, not just the region immediately surrounding the present solution. Unless you are exceptional, you will need help to achieve this aim, in the form of measures to guide your search into profitable areas of possibility that you might otherwise overlook. For this purpose I recommend two types of measures: (a) using *system* to direct your search into many different areas of possibility, and (b) using methods that direct your search into many areas *randomly*. It pays to use both approaches.

Methods of Adding System to Your Search   An excellent method of achieving system is to concentrate on the solution variables one at a time, attempting to generate many possibilities for each. For example, an engineer engaged in the design of an improved system for harvesting apples identified such solution variables as the method of separating apples from the tree, the means of bringing the separator to the apples, method of collecting the separated apples, source of power, and so on. As he concentrated on

Figure **4**   *A page from the engineer's notebook, showing some of the alternative partial solutions he generated for the apple-harvesting problem. For each major solution variable, he classified his ideas in the form of an* alternatives tree. *The use of solution variables to systematize your search can be very beneficial; I urge you to employ them frequently.*

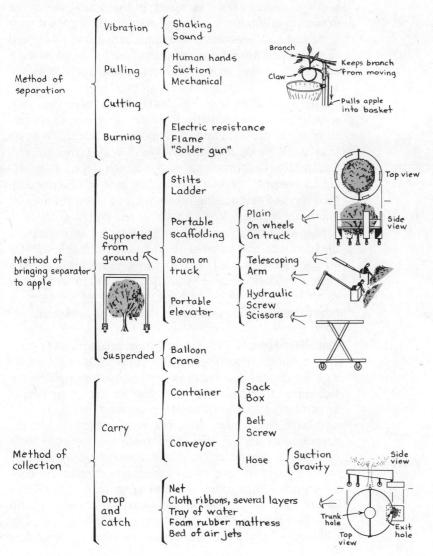

the solution variable "method of separation," he looked first for *basic* methods of separation and then for specific versions of each of these basic alternatives (Figure 4). This is an excellent practice. It minimized the chance that the engineer would overlook a whole block of promising possibilities. He then concentrated on other solution variables in turn, attempting to accumulate as many possibilities for each, always working from the general to the specific.

From now on I will refer to the alternatives for a solution variable as *partial solutions.* In the next phase of the design process the engineer will evaluate these partial solutions and combinations of them, perhaps recombining and re-evaluating numerous times, until he has synthesized a complete solution that is the best combination of partial solutions.

There are other ways of introducing system into your search. You can systematically question various characteristics of the problem and of solutions generated. (Why do it at all? Why do it this way? Why? Why?) You can systematically concentrate on each criterion, trying to generate all possible ways of minimizing cost of construction, then ways of maximizing reliability, and so on through all important criteria. You can be systematic in combining partial solutions, in soliciting suggestions, in examining the literature, and so on. The *alternatives tree* (Figure 4) is an effective means of systematizing your thinking. Any other way of organizing your thoughts and investigations so that a wide range of basically different solutions is brought under consideration is likely to be profitable.

Random Methods   There are some predominantly random methods of getting the mind into what might otherwise be unexplored territory. A notable example is the technique of *brainstorming.* A half dozen or so people assemble for the purpose of generating solutions to a problem. The leader describes the problem; then the participants call out ideas and these are recorded on a blackboard. The objective is to accumulate *many* ideas by creating an atmosphere that encourages everyone to contribute all solutions that come to mind, regardless of how absurd they may seem at the moment. Evaluation and ridicule are forbidden. After some experience with this technique, a group can usually achieve a free-wheeling flow of ideas that yields a surprising number and variety of solutions.

One reason brainstorming is frequently fruitful is because the rapid-fire flow of ideas is repeatedly directing each participant's thoughts into different channels. In terms of the model (Figure 3), each person's mind is buffeted about the solution space in random fashion, forcing "big jumps" to distant points and thus combating the clustering tendency. The probability is high that some of these excursions will enter into profitable areas.

If in a particular design project not enough engineers are directly involved to make brainstorming possible, others can be called in from different projects for an hour or so to contribute their ideas by means of this procedure. This can be very profitable, since a project can benefit from the thinking of a number of engineers while requiring a negligible amount of their time.

Analogies provide another random means of starting your thinking along new avenues. Noting the solutions to analogous problems can be fruitful. For example, if the problem concerns propulsion of a vehicle through water, you might consider how fish propel themselves, how insects propel themselves across water or through air, how worms propel themselves through soil, and how man-made machines are propelled through air. Mental exercises of this type often can lead you to profitable possibilities "by chance."

Terminating Your Search Prematurely   There is a tendency to quit seeking solutions before it is necessary or desirable to stop. This is likely to happen if you become involved prematurely with details or with evaluation of solutions. Therefore:

Don't get bogged down with details sooner than necessary. Suppose you start working out the details of the first "good" idea you have. For all practical purposes your search will end right there. You will be spending time on details when you should be searching for other basically different solutions. Furthermore, preoccupation with the details of one solution severely hampers your ability to think of significantly different ones. Too, if you fall prey to this temptation and later you happen to uncover a superior solution, you will probably be unjustly biased in favor of the solution in which you have already invested so much time on details (and don't think that *you* couldn't be guilty of this!) And, finally, many alternatives can be satisfactorily evaluated while still in a relatively crude state of specification; since most will be rejected, why waste time on detailing them?

Therefore postpone details until they become necessary for decision-making purposes. In fact, it is best to form only solution concepts in this phase of the design process. (A *solution concept* is the essence, the gist, the general nature of a particular solution. Its form may be a rough sketch, a few words, a sentence or two. Thus the ideas shown on page 139 are actually partial solution concepts.)

Avoid premature evaluation; it has the same detrimental effects as premature preoccupation with details. This is the *search* phase of the design process; it is followed by the *decision* phase, in which evaluation of alternatives predominates. Therefore ideas will not go unevaluated but good ideas may go undiscovered if you become preoccupied with evaluation when you should be searching for better solutions.

Don't be too quick to judge possibilities. Most of us have a tendency to dismiss ideas if when conceived they strike us as ridiculous, unworkable, or unprofitable. In so doing, we cast aside some worthy possibilities. But what's the hurry? What seems infeasible now might well evolve into a first-class idea, so keep an open mind. At this stage let every idea remain as a candidate for later evaluation.

This discussion of techniques for improving your inventiveness should be enough to convince you that there are constructive steps you can take. What I have given is only a sample; there certainly are other measures. I hope you will find the time to explore the wealth of literature on this subject and to learn other approaches.

Now is a good time to review page 128. Make sure that you understand the significance of the key phrases in the diagram and get the "big picture" fixed in your mind.

Somewhere along here you should start wondering to yourself, "When do I stop searching and concentrate on evaluating the possibilities I have accumulated?" That's a good question. The idealist might answer, "When you've found the optimum solution." But this could take years, and furthermore you would have a hard time knowing whether you did have the optimum. The pragmatist answers, "When my deadline dictates, or if it's up to me, I use my judgment in deciding when I have reached a point of diminishing return." I'm with him. Since we must use our judgment, differences of opinion and policy do arise. To illustrate the extremes and to help clarify your thinking I pose this question: Would you rather have more alternatives to choose from but spend less time evaluating them, at the risk of not selecting the best, or would you rather spend more time exhaustively evaluating fewer solutions and thus be sure you choose the best from this smaller sample.

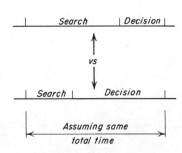

A Plea for Originality

There is dire need for more original thinkers in engineering (in every field for that matter). Too many solutions are the offspring of handbooks or of traditional practices that have little virtue other than longevity. Too few solutions represent original thought. Inertia perpetuates a host of inferior solutions in the world about us, leaving great opportunities for the problem solver who will rely more heavily on his inventiveness.

Your solutions will tend to be unimaginative if you over-rely on the backlog of stock solutions that has accumulated over the years. Relying on this source is tempting, for it is a path of least resistance and provides solutions in which you can be reasonably confident. In general, the more you know about a specialized branch of knowledge, the more stock solutions you are familiar with and the easier it becomes for you to rely too heavily on this source. Often a student confronted by a design problem in a field about which he knows very little complains, "How can I solve this problem? I don't know anything about it." Little does he realize that his ignorance is a big factor in his favor insofar as originality is concerned. He probably will produce some very imaginative designs simply because he is forced to; he knows few or none of the customary solutions and hence can't fall back on them.

Now the point of all this is *not* that you should avoid handbooks or that depth of specialized knowledge is a bad thing. Rather the point is this: draw on the backlog of stock solutions *and* on your capacity for original ones; both have much to contribute. The trick is to be creative in spite of your specialized technical knowledge, without which you cannot get along in modern engineering.

Employers are vitally interested in performance; they don't hire you simply because you are a walking storehouse of factual knowledge. They want you to put it to use, *to create* useful devices, structures, and processes. I mention this partly because at times during your engineering education you may get the impression that originality is not wanted, expected, or appreciated. Although that may be true in certain courses, it is not usually the case in the real world. The fact that there is ample opportunity to exercise your creative powers and to make original contributions should be good news, for many of you will find this aspect of engineering to be the most gratifying part of your work.

Exercises

1 *A plea has been made for more originality in solutions. Start a "scrapbook" of examples of what you believe to be especially original solutions to engineering problems. (Look for radical departures from what was conventional for long periods—for example, the "bouncing ball scheme" of the IBM Selectric typewriter.)*

2 *With a group of students (preferably 8 or 10) generate as many solutions as you can in a 20-minute brainstorming session, for one of the following problems. Someone should record all ideas on the blackboard.*

   (a) *Ways of getting more ideas.*

   (b) *New applications for rechargeable batteries. (Can you supply the manufacturer with 75 uses?)*

   (c) *Product and service ideas for industrial development in an underdeveloped community. (The instructor will provide general information about the community and its resources.)*

   (d) *The "shifting sands" problem along desert highways.*

   (e) *Innovations in residential building construction.*

   (f) *The "wrong-way problem" on dual highways (a "foolproof" system for preventing drivers from entering via an exit).*

3 *For one of the following problems identify the major solution variables; then generate as many partial solutions as you can for each variable. Present them in the form of alternatives trees.*

   (a)   *The English Channel crossing problem.*
   (b)   *An alarm clock for deaf persons.*
   (c)   *Transoceanic communication system.*
   (d)   *Snow removal from city streets.*
   (e)   *Typewriter design.*

Design Projects

1 *An engineering consulting firm has approached a shipbuilding company with an idea. It is to market a fleet of unmanned, automatically guided, submarine cargo vessels. The shipbuilding company believes there is promise in the idea and has commissioned the consulting firm to explore it further. The consultants are to provide a general description of the system they visualize. No detailed specifications are desired at this time. Assume that you are the consulting engineer assigned to prepare this report.*

2 *The reeds (page 7) slide down a channel leading from the feeder device to the final-positioning mechanism. They are randomly oriented as they leave the feeder but must be "paddle first" at the point of assembly. Therefore an orienting device must be interjected between feeder and positioner. Design one. (These reeds are about $1\frac{5}{8} \times \frac{3}{16} \times \frac{1}{20}$ inches.)*

3 *The board of trustees and administrative officers of a university want an improved system for parking autos on the campus. This system should be less disfiguring to the landscape, should accommodate 25% more autos, and should not require a staff member to park more than 100 yards from his place of work. The trustees and administration are convinced that "much can be done with some money and some imaginative thinking."*

THE DESIGN PROCESS: DECISION PHASE

IN the search phase you expand the number and variety of alternative solutions as represented by the upper half of Figure 1. What is needed now is an elimination procedure that reduces these alternatives to the preferred solution, pictured in Figure 1 and described in this chapter.

Initially, eligible solutions are specified only in general terms, perhaps in words or sketches. After the obviously inferior alternatives have been eliminated, usually by relatively quick and crude evaluation procedures, more details are added to the remaining possibilities, which are then evaluated by more refined methods. This multistage screening process continues until the preferred solution emerges. As it progresses, different combinations of partial solutions are evaluated to determine the optimum.

The Tire Mounter—a Case Study An automobile manufacturer had for some time been using a relatively expensive machine to mount tires on wheels before final assembly. Because of an increase in the number of mountings to be made, an engineer was asked to do something to speed up the machine. The result of his efforts is a new and much simpler device (Figure 2). Before he proposed this device to the management he made a thorough investigation and comparison of his solution and the existing one. The results are given in Figure 3, which shows the criteria on which he based his decision, the performance (converted to dollars) of the alternatives with respect to these criteria, and the tabular summary he prepared to facilitate comparison. This case is unusual in that it involves only two major alternatives, but its uncomplicated nature makes it an excellent source of examples for the following discussion.

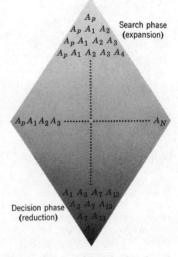

Figure 1 A model of the search and evaluation phases of the design process. The symbol A represents an alternative solution or partial solution.

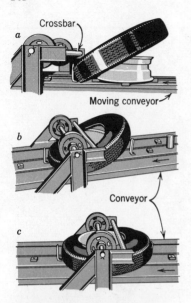

Crossbar

a

Moving conveyor

b

Conveyor

c

Figure 2 (a) The beads of the tire are lubricated to insure that they will slip over the rim. The tire is on the wheel at an angle of approximately 30° to the rim as it approaches the cross bar.

(b) As the tire moves under the cross bar, it is engaged by 2 wheels located parallel to the centerline of the conveyor. These wheels are the only moving parts of the new machine and serve to squeeze the leading beads of the tire into the drop center of the wheel.

(c) The remainder of the tire is forced onto the wheel by the cross bar.

Figure 3 This is a page from the engineer's final report, comparing the performance of his device and that of the one presently used. He predicts that $750 will be required to build, install, and "debug" his device. The costs of an attendant, of maintenance, etc., have also been predicted for the five years that he expects his device will be needed. He has acquired equivalent cost data for the existing machine. He has summarized these costs in this table and (at the bottom) in the form of two ratios that aid readers to appreciate the merit of this $750 investment. It is obvious why the proposed device was readily adopted.

The General Decision-Making Process Although the specifics vary from situation to situation, in almost every instance these four steps must be taken before an intelligent design decision can be reached: (1) criteria must be selected and their relative weights determined; (2) the performance of alternative solutions must be predicted with respect to these criteria; (3) the alternatives must be compared on the basis of these predicted performances; and then (4) a choice must be made.

1. Usually the overriding criterion is the *benefit-cost ratio,* which is the benefit expected from a solution relative to the cost of creating it. For the tire mounter the *benefit* is the saving in operating expenses made possible by the proposed device. The cost is the total expense incurred in building and installing it. For a proposed dam the benefit-cost ratio is

$$\frac{\text{Revenue from power} + \text{conservation advantages} + \text{recreation benefits}}{\text{Cost of land} + \text{construction cost} + \text{operating expense} + \text{hardship to displaced persons}}$$

Knowing that a certain machine will cut his harvesting costs by X dollars a year does not mean much to the fruit grower until he also knows the cost of the machine. Tax-

Criteria		Proposed Device	Present Machine
Investment	Cost to build, install, "debug"	$750	0
Operating expenses (for 5 years)	Attendant	$600	$23,000
	Maintenance	350	900
	Repair	250	750
	Power	0	400
	Total	$1200	$25,050
The unquantifiables	Reliability	Better	
	Safety	Better	

$$\frac{\text{Benefit}}{\text{Cost}} = \frac{\text{Saving in operating expenses}}{\text{Cost of building \& installing}} = \frac{\$25,050 - \$1200}{\$750} = 31.8$$

$$\text{Annual rate of return on investment} = \frac{(\$25,050 - \$1200)/5 \text{ years}}{\$750} = 6.0$$

payers are interested in more than the benefits offered by a proposed public works project; they want to know also what the improvement is going to cost them. The investors in each of these examples want the most for their money—they want to maximize the benefit-cost ratio. And so an engineer rarely describes the benefits attributable to a proposal without quoting in the same sentence the cost of bringing about those benefits.

Ordinarily, to satisfactorily estimate the benefit-cost ratio, a number of subcriteria must be evaluated first. In the aggregate these subcriteria determine the benefit-cost ratio. Some of the subcriteria applied in the design of consumer goods are shown here. As an example, in considering alternative designs for a mechanical toothbrush, the engineers had to evaluate them with respect to these subcriteria before they could predict the overall benefit-cost ratio. Thus they had to evaluate different power sources with respect to cost, reliability, safety, etc., and likewise for different brush actions, mechanisms, materials, and so on.

2. Predicting how well each alternative will fare if it is adopted is the key and most demanding part of the decision-making process. For the proposed tire mounter, the engineer had to *predict* the cost of building it, the number of hours to maintain it, its reliability, and so on. To make these predictions he relied mainly on his judgment and on experiments with a working model of the proposed device.

Of course, predicted performances should be in the same units if they are to be accumulated and compared. By far the most convenient unit is the dollar. Therefore Figure 3 shows predicted performances in dollars for those criteria for which this is feasible. There will always be some criteria —the unquantifiables—which are not easily put into numbers. Note that although it was too expensive and time-consuming to measure safety and reliability *in this case,* the engineer did not ignore them. He judged them in qualitative terms and found that they reinforced the monetary argument for this device.

3. To make an intelligent choice from the alternatives, they must be meaningfully compared with respect to the criteria. When dealing with criteria for which monetary predictions are feasible, the figures are usually tabulated or otherwise aggregated so that costs and benefits can be easily compared, as shown in Figure 3. Incidentally, this example, simplified for our purposes, illustrates one of a number of methods of making economic comparisons of

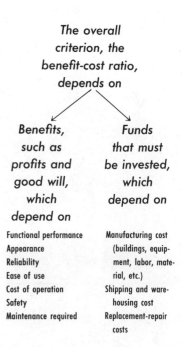

The overall criterion, the benefit-cost ratio, depends on

Benefits, such as profits and good will, which depend on	Funds that must be invested, which depend on
Functional performance	Manufacturing cost
Appearance	(buildings, equipment, labor, material, etc.)
Reliability	
Ease of use	
Cost of operation	Shipping and warehousing cost
Safety	
Maintenance required	Replacement-repair costs

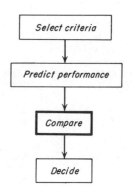

Select criteria

Predict performance

Compare

Decide

engineering alternatives. These procedures derive from a rather extensive body of knowledge of fundamental importance in engineering, generally referred to as *engineering economics.*

Other than outlining the above procedure, not much more can be generalized about decision making in engineering. From problem to problem the particular techniques, skills, and knowledge that you use will vary considerably, as determined by the complexity and competitiveness of the alternatives, the relative importance of the decision, and other circumstances. The decision process ranges from the most elaborate, exhaustive procedures involving much measurement, investigation, prediction, and cost comparison, to the quick, simple, informal judgment.

More on Criteria   A few terms commonly used in engineering discussions of criteria need elaboration. A term used synonomously with benefit-cost ratio is *return on the investment,* meaning the benefit returned by an investment relative to the amount of the investment. Another alternative is *effectiveness-cost ratio.* (Frequently writers and speakers use "cost-benefit" or "cost-effectiveness" when they mean benefit/cost.)

Reliability has a very specific meaning: the probability that the component or system in question will not fail during a *specified period* under *prescribed conditions.* (Example: the reliability of the video tube made by the Zeon Company is 0.95 for one year of operation under "normal" usage, vibration, and temperature conditions.) Reliability is especially important when failure is costly, as in the case of an amplifier in a trans-Atlantic cable under 9000 feet of water. Obviously high reliability is of vital interest to people in the aerospace business.

Operability refers to the ease with which a given design can be operated by human beings. Some computers, automatic washers, cameras, weapons, etc., are relatively easy to operate, require a minimum of learning time, and are difficult to "louse up"; others are quite the opposite. Surely you have encountered both types.

Availability is the proportion of time a machine is usable and thus not "down" for repair, maintenance, or other forms of service. This criterion is especially important when a lot of money is tied up in the machine. If you invest eight million in a commercial airliner, you want availability at a maximum. It is also important when people are highly dependent on the system, as is true of public water supply systems, defense missiles, and elevators in a forty-story building.

Reliability, operability, and availability, along with such criteria as repairability and maintainability, are becoming more important as modern engineering creations become increasingly complex and costly and as we become more dependent on them. A whole production system depends on the tire mounter; a corporation can "go to pieces" if its large computer breaks down for long; and you probably know what happens when the electric power system cuts out for a few hours!

Note that the reliability, operability, and maintainability of a solution—in fact its total cost—depend on its *simplicity,* so much so that I am moved to make a special plea on behalf of this criterion.

A Plea for Simplicity

Among the many solutions to an engineering problem there are some that are relatively complicated; others are refreshingly simple but no less effective than their complex counterparts. The tire mounter provides an excellent example. The existing machine was unnecessarily complicated, it had sixty moving parts and required electricity and compressed air. The new device is in sharp contrast: it has only a few moving parts; there is very little to fail or to maintain. This is good engineering.

As another example, an artificial satellite tends to tumble indefinitely as it orbits; yet for certain applications one side of it must remain pointed toward the earth at all times. This is true of the weather satellite; its camera should always point toward earth. Therefore such a satellite must have an orientation system that prevents it from tumbling and keeps it properly aimed. A long boom attached to the satellite provides a beautifully simple solution to this imposing problem, as explained by Figure 4.

To fully appreciate the virtues of this solution you must know something about the alternatives. You would be amazed at the complicated systems that have been used or proposed to achieve this same purpose. One system uses electronic horizon sensors in conjunction with gas jets.

Figure 4 *The pull of gravity on the
end of the boom closest to the earth is
greatest at that point (the force of
gravity varies with the square of the
distance to the earth's center). This
swings that end toward an imaginary
axis extending from the satellite to the
earth's center. This gravity-induced
torque causes the structure to seek the
desired vertical orientation. In position
A, gravity is in the process of orienting
the still-tumbling satellite. Eventually
the structure reaches and remains in
position B. In space jargon this is
referred to as the gravity-gradient
method of satellite stabilization. Oh, if
only such simplicity prevailed throughout our space program! (Incidentally,
since there is no air resistance, this
structure will act like a pendulum on a
pivot and swing to and fro indefinitely
unless some provision is made to
dampen this oscillation. This is done,
for example, by attaching a tiny weight
to the tip of the boom by means of a
coil spring, so that the weight flops to
and fro out of phase with the boom
and eventually kills the oscillations. It
is possible that by chance the satellite-
boom structure orients itself with the
satellite facing the wrong way, which
would mean that the satellite's cameras
would be forever aimed at outer space.
In this event the boom is retracted,
the satellite is allowed to tumble, then
the boom is re-extended, and this process is repeated until by chance the
structure assumes the desired
orientation.)*

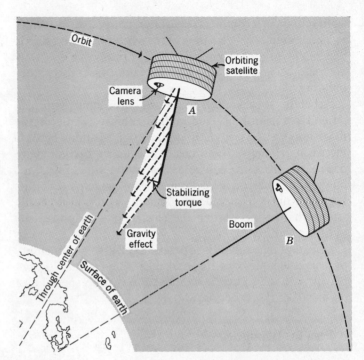

This solution requires compressed gas and electricity (the
gravity gradient method requires no power), and it has
about forty times as many parts, weighs about four times
as much, costs about fifty times more, takes more space,
and is considerably less reliable than the gravity gradient
system. The advantages of the latter system are no small
matters and are due to an engineer's conviction that "there
must be a simpler way."

The stabilizing boom on a satellite may be anywhere
from 25 to several hundred feet long and obviously cannot
protrude during the launch phase. So how does a satellite
grow a boom, say 75 feet long, after it is in orbit? Here, too,
many rather elaborate schemes were considered, but the
solution eventually conceived is remarkably simple; see
Figure 5. Notice some of the virtues of this simple "boom
grower," especially in contrast to the solution that might
have evolved. It is very compact, light, highly reliable, relatively inexpensive and requires very little power. In fact,
the spring action of the metal can provide the energy
needed, so that it can be operated without a motor.

The simplicity of the solutions adopted in these three

case studies is impressive indeed. As usual, these relatively simple creations are economical to manufacture, easy and cheap to use and maintain, and highly reliable. Also, from the point of view of professional pride the solutions that are strikingly simple are the most satisfying. Therefore simplicity is well worth striving for. Do not be content until you have simplified to the maximum extent feasible the mechanisms, circuits, method of operation, maintenance procedures, and other features of your solution. Ordinarily there is a big difference between a workable solution at the time it is conceived and that solution after it has been effectively simplified.

In Chapter 2 I pointed out how a cleverly simplified solution tends to deceive the untrained eye concerning the difficulty of the problem and all the knowledge, skill, and effort that went into the solution. This is certainly true of the three instances cited here. How easy it is to grossly underestimate what underlies this simplicity, especially if you are unaware of the overcomplex solutions that could have evolved (and in fact often do). When a solution of yours has reached this deceptively simple state, you can usually consider your job well done.

A mark of an exceptionally creative person in almost *any* field is the simplicity of his works. Observe the few lines required by a good cartoonist to create the desired effect, the remarkably few brush strokes on a good painting, the few well-chosen words required by the skillful author to express his message clearly, the simple lines in the great works of architecture, or the striking simplicity of the three engineering creations just described. Yes, there is art in engineering and by the simplicity criterion these three solutions are quite artistic.

Often an engineering solution that is especially simple compared to what it accomplishes is described as *elegant*. Since complexity is the inverse of simplicity, elegance is

$$\frac{\text{What a solution accomplishes}}{\text{Its complexity}}$$

The complexity of a solution can often be approximated satisfactorily by counting its component parts (resistors, transistors, gears, cams, etc.), but the ratio itself is difficult to quantify. Nevertheless the elegance concept is a useful one.

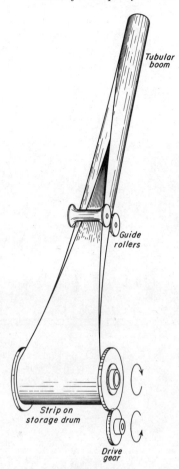

Figure 5   Design concept for "growing" a long tubular boom on an orbiting satellite. A flat metal strip the length of the desired boom is treated so that its normal form is tubular. Then it is rolled up flat onto a spool and stored in the satellite. After the satellite is in orbit a signal actuates a motor that unrolls the strip. As it unfurls, the strip returns to its tubular form and when fully extended is the desired boom. Tubes many hundreds of feet long can be so constructed.

Where Do Technical and Scientific Knowledge Come In?

Could just anyone have thought of the solutions described for the tire-mounting, boom-growing, and satellite-stabilizing problems? If not, what special knowledge was required? A clever person without an engineering education could have thought of the solution concept underlying the new tire mounter, but he would be hard put to specify the mechanical details of a *workable* device. In the case of the boom grower, a good machinist could have conceived of the solution concept, but the engineer who developed the functioning mechanism had to have special knowledge of metals, heat treating, stresses, mechanisms, and cantilever beams. Only a person familiar with principles of mechanics, gravitational phenomena, oscillatory systems, and the like would be equipped to conceive—let alone develop —the gravity gradient idea.

The nub of the solution in each of these instances is pure invention, a product of an engineer's ingenuity. But without special technical and scientific knowledge it would have been virtually impossible for a person to convert the basic idea to a workable solution in any of these cases. Although invention is a necessary and very important part of engineering design, it is hardly sufficient.

This analysis also highlights a major difference between modern engineers and men like Edison, Whitney, Watt, and others of preceding periods. All are inventors, but there is a big difference in the amount and type of knowledge utilized by the two groups.

Exercises

1   *Some elegant solutions to engineering problems have been described. Identify five more examples from the world about you. Select what you believe to be the most elegant of these five, and submit it for competition with similar entries from other class members. A list of these will be distributed, and each student will rank the creations on the list. Assign the number 1 to the entry you judge to be most elegant, 2 to the second-ranked one, etc. The results of the voting will be tabulated and announced. (I am interested in learning of the results.)*

THE DESIGN PROCESS:
SPECIFICATION OF A SOLUTION;
THE DESIGN CYCLE

THE input to this phase is the chosen solution, some of it in the form of rough sketches, notes, computations, and the like, and much of it still in your head. Besides being incomplete, this material is disorganized and hardly in shape for presentation to superiors or clients.

It remains for you to describe the physical attributes and performance characteristics of your proposed solution in sufficient detail so that persons who must approve it, those charged with its construction, and those who will operate and maintain it can satisfactorily fulfill their responsibilities. The fact that someone other than you ordinarily constructs, operates, and services your creations makes it especially important that you carefully document and effectively communicate them.

The output of this phase usually consists of engineering drawings, a written report, and possibly a three-dimensional iconic model. The first of these communication media, often called the prints, are carefully prepared, detailed, dimensioned drawings of the solution.

The second medium, the engineering report, is usually a rather formal document describing your proposal in words, diagrams, and sketches. This report also describes the performance of your solution and provides a thorough evaluation of it. It is through such reports that your ability to express yourself becomes apparent to the people you want most to impress favorably.

Occasionally you will supplement the prints and your report with a physical model (page 52). This is an effective communication device and an aid in gaining acceptance of your proposal by superiors, clients, and the public.

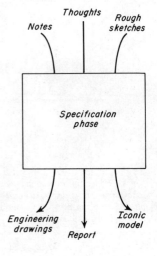

This phase of the design process is likely to involve considerable detail. Draftsmen and other technical assistants can relieve you of some of the burden, but in general *you* must specify the types and properties of materials from which your creation is to be made, as well as dimensions, methods of fastening, tolerances, and similar essential details.

The Design Process in Retrospect

Probably you are the victim of certain habits of thought that interfere with your ability to solve problems effectively, like the tendency to bypass problem definition. If you are to overcome these bad habits, *you must work at it.* Conscientious disciplining of the mind is required. The effort is worthwhile, though, for the payoff is superior results in solving professional *and* personal problems. I urge you to devote careful attention to your design technique. Study and apply the process summarized in Figure 1 until it becomes natural to you. Examine your approach and indulge in some constructive self-criticism. After you have had experience, you may wish to modify the procedure I have recommended to better suit your needs and preferences. Fine; this means you are paying attention to your design methods—exactly what I want.

A survey of progressive engineering practitioners and

Figure 1   Phases of the design process, showing the probable inputs and outputs of each phase.

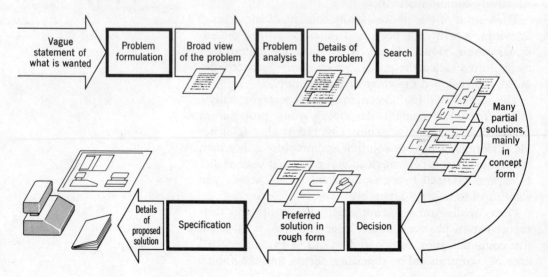

educators indicates a general belief in the existence of a particular design procedure that *in the long run* yields superior results, both in quality of solutions *and* in the cost of arriving at them. It is true that even the most inferior problem-solving approach will occasionally yield a commendable solution, for an element of chance is involved in the generation of ideas. Furthermore, use of an optimum approach will not guarantee that the final solutions to all problems will be superior to what inferior procedures might have produced. The difference lies in the *probability* of superior results, so that the payoff is in your long-run performance.

Although there is general agreement on the existence of an optimum design procedure and on its main characteristics, authorities differ on specifics. For example, all authors urge that you begin your attack on a problem with a careful definition of the problem. Here I differ from the majority by recommending a two-stage problem definition —the "big picture" first, then the details. Some authors have nothing specific to say about how to define a problem; others recommend a pattern and a nomenclature. I recommend a specific pattern and have cited good reasons for doing so. I mention these differences so that you will be prepared for them, should you do further reading on the subject of design—and this I urge.

The design process is a series of stages in the evolution of a solution to a problem. The purpose of each phase is different; also, the type of problem-solving activity that predominates in each is different. However, these stages do not have sharply defined boundaries, nor are they the orderly progression of distinct steps for which the idealist would hope. There is a fuzziness as the emphasis shifts from one phase to the next. Occasionally, solutions will occur to you while problem definition is the predominant activity; during the search phase you may decide to reformulate the problem. Similarly, it is impossible not to do some evaluating in the search phase. Chance plays a significant role in this process; new information, new ideas are uncovered unpredictably, adverse consequences discovered, blind alleys encountered. All of these result in irregularity and occasional backtracking, as pictured in Figure 2.

In some cases the design process deals only with the general characteristics of the solution, temporarily treating

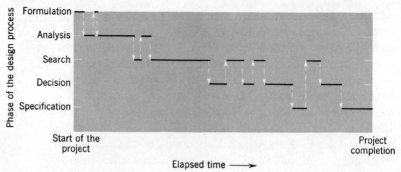

Start of the
project

Project
completion

Elapsed time ⟶

*Figure 2   A graphic representation of
the distribution of an engineer's time
over phases of the design process
during a project. Such graphs for different projects would vary radically;
some would be relatively orderly;
whereas others would be much more
irregular than this one.*

subsystems and components as black boxes. This is sometimes done in a feasibility study, where details are carried
only far enough to enable the engineer to satisfactorily
predict the costs of developing and producing the device
and to forecast consumer acceptance. Detailed design is
contingent on favorable predictions.

In the design of large-scale systems a similar procedure
is employed because of the complexity of the total under-
taking. A communications satellite has numerous major
subsystems, each with hundreds or thousands of components; many engineers are involved. In the design of such
a complicated device the general features of each subsystem are specified with little concern for component details.
After the broader aspects of the overall system are tentatively set, detailed design of subsystems and components
begins. A team of engineers is assigned to each major subsystem, each team being given certain inputs, outputs,
restrictions on size and weight, etc., that were fixed in the
overall-system design phase. Design of the overall system
and design of its subsystems are obviously closely interrelated, as are the activities of the different teams.

The Recurring Question of Economic Feasibility   There
are remarkably few things that man cannot achieve given
sufficient time and money. Whether a device *can* be developed for a given purpose is seldom the question. It can
almost certainly be done *if* someone is willing to pay the
price. The real question usually is, can a solution be developed profitably? Therefore in most instances technical
feasibility is not an issue; economic feasibility definitely is.

Implicit in the inception of a design project is the
assumption that a solution to the problem is economically

feasible, that investment of engineering and other resources to develop a solution will be repaid with a profit. Before starting to design a machine for assembling reed switches, the engineers deliberated over the prospects of a profitable result. The decision to proceed was based on their judgment, on past experience in similar ventures, and on a willingness to accept a certain risk.

From the moment a project is begun, the hypothesis that a profitable solution will evolve is under test. This question is being asked repeatedly, although not always explicitly, through the design process, "On the basis of what has been learned thus far in the project, are the prospects of the venture yielding a satisfactory return on what apparently must be invested sufficiently high to justify continuing?" Thus, at any time from its inception to the specification stage, a project is subject to termination if the information accumulated indicates that a profitable solution probably will not be found under the current state of technology.

Of course at the outset of a design project relatively little is known about the probable final solution, so that at this stage there are many uncertainties and thus a sizable risk of being mistaken in assuming that the venture is economically feasible. As the project progresses, alternative solutions evolve, possibilities are appraised, and other information on which to base an answer to the ever-present profitability question becomes more substantial. Therefore the risk of making the wrong decision is maximum at the start of a project and decreases thereafter as progress is made and evidence accumulates.

Making recommendations concerning economic feasibility of proposed projects is a very important part of an engineer's work. Such decisions are far from simple, and yet, the engineer who acquires a record of blunders in this respect is not regarded favorably.

The Design Cycle

Your task seldom terminates with specification of a solution; responsibility ordinarily extends to gaining acceptance of your design, overseeing its installation and initial use, observing and evaluating it in operation, and deciding (or helping to decide) when redesign is advisable. These functions complete the cycle diagrammed in Figure 3.

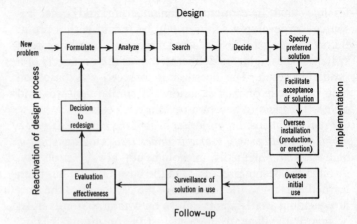

Figure **3** *The design cycle.*

Implementing the Solution   Do not assume that your solutions will automatically be adopted, properly constructed, and used as intended. Many things can go wrong, and measures must be taken to prevent them from doing so, between the time you specify a solution and the time it is an accomplished fact.

For example, measures are needed to ensure that your solution will be accepted by the appropriate people. Engineers often begin their careers with the mistaken impression that, if their proposals are technically and economically superior, they will naturally be adopted. But engineering is ordinarily a staff function in an organization, so that engineers issue recommendations, not commands. This, plus the fact that differences of opinion do arise, makes it imperative that you devote careful attention to the matter of gaining acceptance.

Young engineers are inclined to become discouraged after several of their proposals have been rejected. They are inclined to blame other people, the organization, anything or anyone but themselves. But the chances are they have underestimated the need for effective presentation of their proposals, for selling others on the worth of their ideas, for a certain amount of realistic compromise of some features of their proposed designs, and for careful planning to minimize resistance to change.

Follow-up   Periodic monitoring of your solutions in use is especially valuable as a means of improving future designs. Rare is the engineer who cannot benefit by observing his creations in the field.

Reactivation of the Design Process　Periodic evaluation of solutions in use also provides a basis for deciding when to redesign. No solution to a practical problem remains superior indefinitely. Better methods are discovered, new demands arise, new knowledge accumulates, conditions change, and physical depreciation occurs. Consequently, a point is reached in the life of a design at which it is profitable to seek a better solution. An engineering department can intelligently decide when to engage in redesign only if the current solutions to problems within its realm are periodically appraised.

The design cycle is complete when, after a solution to a problem has been devised and used over a period of years, it appears that redesign will prove profitable, and the process of designing a superior solution is again initiated.

The Range of Engineering Design Problems

Two engineering tasks will be described to illustrate the broad applicability of the design process. Task A involves the design of a device that will directly convert the spoken word to printed form. The input will be an oral message; the output must be a record of that message on paper. Such a device obviously has commercial value.

Task B also involves design. An engineer is employed by a company that manufactures electrical equipment, such as motors and transformers, and assembles these components into power systems designed to suit the unique needs of individual customers. Most of these customers are factories, refineries, printing plants, and the like. One potential customer, a paper-manufacturing company planning to construct a new plant, has requested the engineer to familiarize himself with the paper-making process, to make a thorough investigation of the company's needs, and to then design a complete power system adapted to the process. In doing so, the engineer naturally will rely heavily on components manufactured by his employer. He will submit his design along with the price to the potential customer. If his system is purchased, he will oversee its installation and remain in contact with the job until the new system is running smoothly.

Task A is concerned primarily with the *conception* of a new device. There is a minimum of past experience on which to rely; much original thought is required; some research may be necessary.

Task B is concerned mainly with the *application* of existing devices and components to the satisfaction of the needs of specific customers. For this type of work the engineer can rely on a large backlog of experience in designing such systems in similar situations. Although each situation that he encounters is different to a degree, the design work involved is hardly what would be considered pioneering.

Whereas the main challenge of task A stems from the pioneering nature of the work, the main challenge of task B lies in acquiring a thorough understanding of the requirements peculiar to a specific customer. The main reward offered by task A is the opportunity to contribute something basically new for mankind's benefit; the main reward from task B is the opportunity to serve a customer directly and to observe the results and the satisfaction experienced by him.

These two tasks lie close to the extremes of a wide range of types of design activity (Figure 4). Along this spectrum are such tasks as design of an interplanetary transport, an automatic hair cutter, an accident-proof highway, a factory, an automobile, a cement plant, a bridge—you name it. *The solution of each of these problems requires application of the design process.*

Primarily CONCEPTION of devices, structures, and processes.
Activity at this extreme is highly technical, involving much innovation and exploration.
Experience to guide solution of problems is a minimum.

Range of engineering design activities

Primarily APPLICATION of devices, structures, and processes.
Activity at this extreme involves a minimum of technological innovation.
Experience to guide solution of problems is a maximum.

Figure 4   Engineering design covers a broad spectrum of types of activity suiting many interests and abilities.

13

OPTIMIZING YOUR
PROBLEM-SOLVING METHODS

IN Chapter 6 the concept of an optimum was applied to the solutions to engineering problems. The concept is just as applicable to the methods the engineer employs in arriving at those solutions, for example, the measurement systems, the computational methods, the models, and the number and types of technicians he uses.

For instance, there is an optimum degree to which a model should be refined. Over the long run the errors in a model's predictions cost something. This is so because mistakes, failures, accidents, repairs, and alterations result when decisions are based on erratic predictions or because high safety factors are necessary to protect against such adverse occurrences (e.g., making a beam twice the size that an equation predicts is necessary).

The designers of a chemical plant are relying on an analog model to make predictions on which to base their design. If, after the plant has been built, there is a small disagreement between performance predicted and performance experienced, it is of little practical consequence and is accepted as inevitable. In this instance the cost of the lack of correlation between predicted and actual results is negligible, as indicated by point 1 on curve *a*. A larger discrepancy, however, may well result in some wrong design decisions that are discovered after the plant is built and for which the penalty is costly alterations. In this case the cost of the lack of correlation may be somewhere around point 2. A still larger lack of correlation can result in something much more costly—an explosion, the cost of which may be in the area of point 3. A similar situation exists for all models. This curve indicates what generally happens to the cost of errors in a model's predictions as

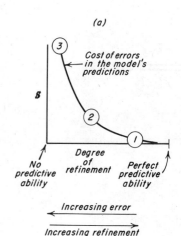

(a)

Cost of errors in the model's predictions

$ \$ $

Degree of refinement

No predictive ability

Perfect predictive ability

Increasing error

Increasing refinement

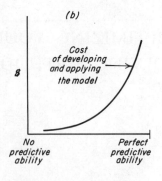

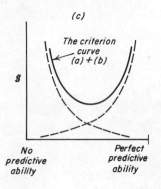

the engineer refines the model and reduces the errors. The cost declines but at a decreasing rate, until a point is eventually reached at which additional refinements are of negligible benefit and not worth striving for.

There is another good reason for not attempting to refine a model to the point where curve *a* levels off. The cost of developing and applying a model builds up as indicated by curve *b*. This cost rises at an increasing rate because, as further efforts are expended to reduce the error in the model's predictions, additional improvements become increasingly difficult and time-consuming to achieve. This is generally true.

The optimum degree of refinement, then, is the point at which the sum of these two costs is a minimum; see curve *c*. It is uneconomical to attempt to refine a model beyond this point.

The situation summarized by curve *c* is not unique to modeling. It prevails for measurement systems, information-search methods, and most tools, techniques, and procedures you employ in your work. In each case there is an optimum degree of refinement simply because the manpower and other resources devoted to such efforts are valuable for other purposes and because continued improvements bring diminishing returns. In fact, for the same basic reasons there is an optimum number of man-hours to devote to the solving of a problem.

Selecting a "Tool for the Job"

There you are at your desk laboring over some problem, and first thing you know you have a system of simultaneous equations to solve. Hopefully, you do not *automatically* solve it by *the* way your professor taught you in college; rather, you pause and think . . . substitution method, subtraction method, trial and error, graphical means, determinants, computer, others? Then, to select the best of these possibilities you consider the probable amount of time required for each method, the cost of the equipment used, and the accuracy obtained. On the basis of these criteria you judge which alternative best suits this situation.

Or suppose you have a pile of figures on your desk, each of which must be squared, added, averaged, etc. No, the computer is not automatically *the* way out. Quite possibly the cheapest and quickest way is the ordinary pencil and paper method, or slide rule, or desk calculator, or scanning the figures and estimating the quantities you need. Or maybe you should turn the task over to a clerk. The point is that there *are* alternatives, and you should consider them before you perform the task at hand.

You Name It—There *Are* Alternatives   These case studies of alternative problem-solving methods illustrate what is true of just about every technique, procedure, and device you will employ. It holds whether you are measuring, predicting, computing, communicating, or whatever. And since no one of the alternatives is optimum in all situations, it behooves you to consider the choices and select intelligently before you tackle a given task. But let me level with you. Engineering schools by and large do not do a good job in encouraging you, let alone teaching you how, to do this. So it is up to me through this book, to a few professors who do emphasize this skill, and *to you mainly,* to develop your ability to optimize your problem-solving methods.

How Not to Select   Partly for lack of time, you are not shown all acceptable ways of solving a given type of problem in your college courses. But don't let the fact that you are taught only one way mislead you; you may safely assume that alternatives worth considering do exist in each instance. Furthermore, in college it is probably wisest to solve each type of problem as instructed, but after graduation you should explore the alternatives and use them when appropriate. Don't go on for years solving a particular type of problem in a given way just because it's *the* method you learned in college. (There is a good chance that it is fancier and more time-consuming than necessary for a certain proportion of cases.)

There are other undesirable reasons for always solving a type of problem in a certain way. One is habit, which of course is the path of least resistance. Often it is much easier to blindly and automatically follow the same procedure each time a type of problem arises than it is to stop and reason, consider, and perhaps change. And don't think that engineering practice is devoid of fads. Nor are all practitioners above solving a problem by an oversophisticated procedure in order to impress their professional colleagues. Surely some tasks are handled by computer, when simpler and cheaper methods would suffice, simply because this is the popular thing and makes a good impression. (But I hasten to add that this is not true just in engineering!)

How Should You Select? *First,* seek out the alternatives available to you for the task at hand, *then* apply sound criteria and select the approach that is optimum in that particular situation. What I mean by sound criteria is illustrated by the following example.

Suppose that you are designing a superterminal to handle several thousand commuter buses per day. In particular, you wish to predict the capacity, queues, conflicts for bays, and general effectiveness of different designs under consideration. For this purpose you can call on one or more of these major methods of prediction: your judgment, a mathematical model, simulation, or experiments with the real thing. In choosing a method for this job you should apply these criteria:

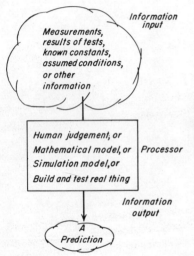

Information input

Measurements, results of tests, known constants, assumed conditions, or other information

Human judgement, or Mathematical model, or Simulation model, or Build and test real thing

Processor

Information output

A Prediction

- The cost of making the prediction, which depends on the man-hours required and the facilities used. This is the obvious part.
- The cost of error in the predictions, which depends on the magnitude of this error and on its consequence in the specific situation. In one situation a large error is of no consequence; in the next, the same error could result in extensive loss of life and property.
- The absolute amount of time required to make the predictions. A judgment can be made in a few seconds or minutes; building a full-scale prototype to test it under actual conditions can delay a project weeks or months.

Figure 1 indicates how the major prediction methods stack up with respect to these criteria. None of these alternatives is best for all cases; each has its place. In some situations judgment is the method to use; it can be erratic, but sometimes error is relatively inconsequential. Furthermore judgment is quick and cheap. In other cases mathematical models or simulations are the best choices. Constructing each design alternative and testing it under real conditions is the best method under some circumstances, even though

Prediction method / Criterion	Judgment	Mathematical model	Simulation	Build it and try it under actual conditions
Cost of making the prediction	Generally increasing ⟶			
Error	⟵ Generally increasing			
Time required	Generally increasing ⟶			

this procedure is usually very expensive and time consuming (obviously and prohibitively so in situations like the bus terminal example).

Elegant "Tools" So far I have talked about optimization, benefit-cost ratio, alternatives, and criteria as they apply to your problem-solving methods. Other matters discussed earlier in the book apply here too. One that is particularly appropriate is the concept of elegance (page 153). Observe that my pleas for simplicity and elegance apply to the methods you employ in arriving at solutions as well as to the solutions themselves. In this sense elegance is the utility of a "tool" relative to its complexity. There are elegant computer programs and not-so-elegant ones, and the same is true for methods of stress analysis, computational procedures, measurement methods, and so on. Although you may think that *you* would never use an unnecessarily complex technique or procedure in the course of solving a problem, this point *is* worth mentioning. By all means use computers or high-powered mathematics when they are the best means for the task at hand, but don't be like the fellow who gets a bulldozer for a two-minute pick-and-shovel job.

Figure 1   Some generalizations about major methods of prediction. These are generalizations; surely there are exceptions. Note that I have not ranked the methods with respect to cost of error, which cannot be evaluated until we start discussing specific cases. In general, when faced by a specific prediction task, I consider these alternatives starting from the left and moving to the right. I check first to see whether my judgment is adequate; if not, I investigate mathematical models; and so on.

Exercises

1   *After reading this chapter you should have something intelligent to say about "the right procedure versus the best procedure" for solving a given problem. Prepare a brief essay on the topic.*

2   *There is an optimum period of time (calendar days) for completion of a project. Draw the curves (general, no scales necessary) portraying major conflicting costs and total cost. What relevance does this have to the man-on-the-moon and supersonic transport projects, both of which are crash programs?*

3   *From one of your present courses, choose a particular type of task—computational, prediction, measurement, etc.—and describe as many alternative methods of doing the job as you can. Identify the criteria you would use to choose the best procedure for a particular case.*

14

ENGINEERING AND SOCIETY

TOOLS, machines, and structures often have profound effects on the lives of men; many have been closely intertwined with major political, social, military, and economic events in history. Visualize, for example, the commercial, cultural, and political interactions between nations that must have occurred when vessels capable of traveling the high seas were developed; the increase in agricultural productivity that resulted when the iron plow replaced its wooden predecessor; the effect of the printing press on the preservation and dissemination of knowledge; the military and political impact of a major new weapon like the atom bomb; the social and economic impact of the automobile. These are not the only effects of these contrivances, nor are these the only contrivances that have had important consequences for mankind.

The devices, structures, and processes created by present-day engineers are no different in this respect. These creations significantly affect man's physical comfort and safety; his mobility; the ease with which he communicates; the education he needs; his life span; the hours, content, physical demands, and stability of his job; his leisure activities; his physical environment. In fact, our economic, social, political, and military systems are strongly influenced by and dependent on the creations of engineers.

Take one of these—physical environment. The densely populated urban area with its tall buildings, conveniences, noise, congestion, and polluted air; highways; man-made bodies of water—these are some aspects of your outdoor environment for which engineering is primarily responsible. And note that you can have an indoor environment that remains remarkably close to the ideal, regardless of outdoor lighting, temperature, and humidity conditions. The effects of this man-made environment are more than

Cotton gin

Airplane

Electric motor

Television

Musket

Elevator

Gas stove

Computer

Radar

Linotype machine

Steam engine

Figure **1**

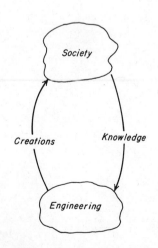

physical. Remember that a person's behavior is determined partly by his environment, and so engineers through their environment-changing creations indirectly affect human behavior.

The main purpose of this discussion is to make you *"impact conscious"*—alert to the far-reaching effects of engineering creations. *Anything you create as an engineer is bound to affect people, probably many persons in numerous ways.* To develop your impact consciousness, I urge you to devote at least a few minutes to Figure 1, in which I have listed a few of the hundreds of engineering creations that have significantly affected the lives of men. I suggest that you reflect for a moment on each one, visualizing the effects you know or can readily surmise that it has had on people.

Sensitivity to the impact of your solutions is important for two reasons. People are *directly* involved with your creations, as users, operators, and maintainers. *People pilot planes, people ride transit systems, people service automobiles, people operate machines in factories. Especially by the effectiveness with which your designs satisfy these people will you be judged as an engineer.*

But jet aircraft make noise that irritates the residents of many communities; the introduction of machines into mines has affected the miners, their families, whole communities; the Verrazzano-Narrows bridge certainly means more to the residents of Staten Island (previously a relatively isolated community) than just a more convenient way of getting to other parts of New York City; construction of a factory to manufacture the desalinator (page 10) in a semirural area will certainly affect that community through new jobs, increased traffic, and "local inflation"; a tunnel or bridge across the English Channel will significantly increase the interaction between peoples of several nations. In such ways as these, engineering creations *indirectly* affect people. *Also on the basis of the side effects—good and bad—of your solutions will you be judged as an engineer.*

If your creations are to serve people well and you are therefore to be judged favorably in both these respects, you must "know" the society your solutions affect and *care* about its well-being (Figure 2).

Get to Know Man—His Individual and Social Behavior
What people need, prefer, and will tolerate should strongly influence the features of your designs. If you are designing

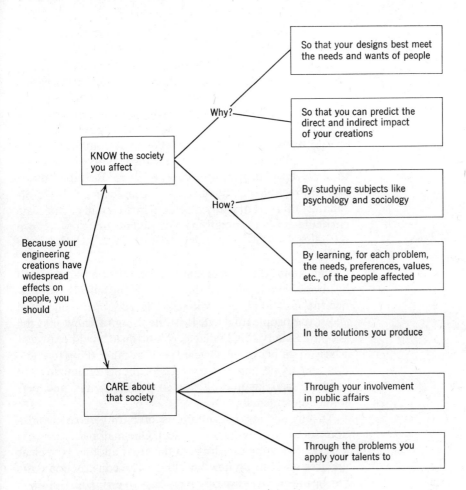

So that your designs best meet the needs and wants of people

So that you can predict the direct and indirect impact of your creations

By studying subjects like psychology and sociology

By learning, for each problem, the needs, preferences, values, etc., of the people affected

In the solutions you produce

Through your involvement in public affairs

Through the problems you apply your talents to

Why?

How?

KNOW the society you affect

Because your engineering creations have widespread effects on people, you should

CARE about that society

Figure **2**   *A summary of Chapter 14.*

a new high-speed transit system for a community and you seek to satisfy the maximum number of people, you had better know what their needs and wants are. Where do people want to travel, and when? How much importance do they attach to privacy as opposed to pickup frequency as opposed to comfort as opposed to other criteria? Are enough people willing to give up the privacy and independence afforded by automobiles to make a new high-speed transit system feasible? How would they react to overhead train-carrying structures passing through suburban areas? How much noise will they tolerate? Finding the answers to these questions is part of getting to know the people who will be directly and indirectly affected by the system. While you are designing it you will use this information to predict how people will react to alternatives you are considering (e.g., overhead versus underground, or individual cars versus trains) in an attempt to maximize satisfaction and minimize opposition.

If you are to predict the full economic, social, cultural, and political impact of your alternative solutions, you had best begin learning about these matters in college. Courses in economics, psychology, sociology, government, and the humanities are recommended for this purpose. Also, you should study in depth the impact of some contrivances like those listed earlier. It will be a profitable experience for you to analyze the impact of several of these, including direct and indirect effects, adverse as well as beneficial. Unfortunately this sort of study is seldom included in engineering curricula, so it is something you should do on your own initiative.

Caring About the Well-Being of Man   Hopefully you will be more than aware of the far-reaching effects of engineering creations; you will also be genuinely concerned about the people affected. It is one thing to know that the people of many small villages will be dislocated because of a large dam being constructed. It is another thing to know this *and* to be actively concerned about minimizing the hardship and facilitating the economic, social, and personal adjustments.

An engineering consultant is preparing recommendations for an urban renewal project. One major alternative is to restore existing dwellings in the area; another is to clear the area and put up new buildings. The construction costs for the alternatives are relatively easy to estimate. But there are other factors to consider—the social consequences, which are quite different for the choices available. I like to think that this engineer is considering these social side effects, along with the tangible costs, as he arrives at his recommendations. I also hope that, regardless of the plan he proposes, he will include in his recommendations suggestions for minimizing the social costs and maximizing the social benefits of this project.

Engineers have been criticized for showing insufficient concern for the full implications of their creations, especially the long-term, indirect effects. Much of this criticism is misdirected—when an automobile company ups the horsepower of its cars beyond all reason, the engineers who designed the engine are at fault only for following the orders of executives who made this decision. However, enough of such criticism is justified to warrant serious consideration of the pleas made in this chapter.

You need not rely on criticism of past mistakes to motivate your concern for these matters. After all, engineers *are* as much creators of social change as of physical change; you *should* know and care about the beneficiaries and the victims of your creations. It is quite appropriate that your designs closely match human needs and wants, that you intelligently predict the full impact of solutions you consider, that you effectively minimize the social disadvantages and maximize the social benefits of your creations, and that you balk when pressured to specify a solution that increases profits at the expense of public safety.

Incidentally, there is more than one way you can express your social concern (Figure 2). You can also show it through your involvement in public affairs, which takes a variety of forms ranging from office holding to the informal activities of the concerned and informed citizen. *There is dire need for all forms of such involvement by engineers.*

Public officials make many difficult decisions of an engineering nature involving the bulk of taxpayers' money, decisions concerning dams, highways, supersonic aircraft, ballistic missile systems, high-speed trains, moon exploration, computer systems, water supplies, pollution abatement, traffic control systems, weather satellites, and so on. Since there are many engineering alternatives and public funds *are* limited, government leaders must decide which projects can be supported and which opportunities must be foregone or postponed. Military decisions have also become technically complex. They involve sophisticated weapons systems, large computer complexes, surveilance networks consisting of radar, satellites, and special aircraft, elaborate communication systems, huge construction projects, and the like. These too are primarily engineering problems, involving many alternatives and very difficult decisions, and in these respects the military and civil situations are similar. What is different between the two is that the military establishment employs sufficient numbers of engineers, directly or indirectly, to assist it in making such decisions, but their counterparts are lacking in civil administrations and legislatures. Government officials need help in the form of more engineers in public service and more engineers who contribute their specialized knowledge and skills as active, concerned citizens.

There are shockingly few engineers in government outside of civil service. The number in elected offices is negligible. In view of the increasing technical content of legislative, administrative, and legal decisions at all levels of government, the number is certainly disproportionately small. The need is obvious; the opportunities are there.

Another way that engineers can contribute is as concerned citizens. The contributions that they can make as members of school boards, urban renewal commissions, industrial development authorities, and similar bodies are obvious. Not so obvious is the need for engineers to communicate to the general public the capabilities of modern engineering, the purposes for which taxes are being used in this area, and the alternative opportunities that exist. I wonder how many voters know approximately how much money the government is sinking into the development of a commercial supersonic transport? Or how many are even aware that any tax money is being used for this purpose? How many know that a VTOL plane is feasible and in the prototype stage, and that this project receives only a tiny fraction of the money allotted to the supersonic transport? Do the people really know what a manned exploration of the moon will ultimately cost them? Do they realize how much more it is costing as a crash program? Are they aware that the moon could be explored at much less cost by machines and instruments controlled from the earth?

How many, if they knew, would rather that this money be spent on better educational facilities or larger and more reliable water supplies or on any of hundreds of other alternatives? In the past, most people knew what their taxes were going for and what the few alternatives were, for example, build roads, collect rubbish, provide fire protection, buy rifles for the army. But now the situation is quite different—there are many more alternatives, these are more complex, and many of them are less obvious and in fact unknown to the masses. Right now the majority of voters have only a vague idea of how their money is being used in the areas of science and engineering, and they know even less about the many alternative ways this money could be employed. They should be informed so that they can express their preferences to public officials. Therefore I urge that engineers, by whatever means possible, do their share to help educate the public in these matters. This *is* a serious situation, for as time passes more and more multimillion-dollar decisions are being made by a handful of people while the electorate becomes ever further out of touch with what is going on.

There is still another way for engineer-citizens to participate in public affairs: by *taking a stand* on critical public issues and *speaking out*. For example, as you may have gathered, the supersonic transport is not high on everyone's priority list for government spending, relative to development of faster and safer surface transportation systems, the VTOLs, and numerous other alternatives. I was pleased when the editor of an engineering magazine wrote:

> Given the fact that fewer than 15 per cent of our U.S. population has ever been up in an airplane, and that perhaps only 3 per cent of them travel regularly by air, it seems to me that Everyman ought to be about due for a break. Take the much-debated supersonic transport plane. As a broad application of our latest technology, it has few peers. But although it may bring us much in national prestige, what we really need is better mass transit. Moreover, what the supersonic transport will do to urban and suburban living does not bear thinking about.*

Such expressions give me pleasure because I want more engineers to make their voices heard, whether I agree or not (although in this instance I'm with him). All too often, though, I read or hear complaints like this:

> Recently a controversy erupted over a plan for a highway along Chicago's lake front. It is interesting to note that hundreds of individuals and a score of organizations expressed themselves—offering comments, requesting changes, or seeking the opportunity to participate in the planning of such an improvement. I cannot recall that a single engineer (out of government) spoke up, pro or con. And to my knowledge, no local engineering society expressed interest. Not so the other professions! Architects and planners were free with their opinions, some quite voluble.†

Take a stand against proposals that will pollute the atmosphere, contaminate the streams, endanger the public, disfigure the countryside, or otherwise strike you as unwise, unethical, unaesthetic, or otherwise undesirable. Speak out!

*Ford Park, "In Our Opinion," INTERNATIONAL SCIENCE AND TECHNOLOGY, October 1965.
† John G. Duba, "Interprofessional Relationships in Environmental Design," CIVIL ENGINEERING, February 1966.

As Figure 2 indicates, a third outlet for your social concern involves the particular problems you choose to solve. For instance, here is a type of problem worthy of your talents and which you may not have thought much about. The VTOL plane, basic processes for converting large volumes of salt water to drinkable form, the artificial kidney, and a host of other potentially significant creations are going through a critical stage in their evolution. They, like most engineering contrivances, must be converted from basic ideas to forms that are technically and economically "fit for the masses." This is the time when ability to simplify, perseverance, and concern for Everyman are especially important. Atomic power is a good case in point; about twenty years of technical and economic improvement effort was required to get it to the point where it can compete with coal and hydro power. Or take the artificial kidney. It has been invented, but a big challenge remains: to get the cost down so that no one need die because he cannot afford such a machine or because there are not enough available. True, this type of engineering work does not offer the glamor associated with the conception of a new device like the kidney machine; but it certainly is no less important to mankind. This and numerous other high-potential concepts depend on the interest of many good engineers.

Plenty of other unsatisfactorily solved problems need your talents and are worthy of them, and the next chapter describes many of these. But there are also problems on which respectable engineers should not waste their time and talents. What you choose to do in this respect depends on your personal values and aspirations. Suffice it to say that the profession needs men who are concerned about the *why and the worth* of the problems put before them, who have the conviction to say "No!" when a task is unworthy of them, who balk when what is expected of them is unprofessional. There will always be room for engineers of this stature.

Exercises

1   *Sir Eric Ashby once wrote that a good engineer can "weave his technology into the fabric of society." What do you suppose he meant by that statement?*

2 *Prepare a paper that analyzes the economic, social, political, military, and cultural impact of one of the creations named in Figure 1, page 170.*

3 *An engineer believes that certain features of a pipeline his company is about to install are unsafe. What should he do about it? (This is a controversial issue, so there is no such thing as THE answer. But you should give very careful thought to what YOU would do under the circumstances.)*

4 *The Aswan Dam on Egypt's Nile River has had far-reaching effects. Analyze the impact of this structure and the measures that engineers have taken to minimize the adverse effects.*

5 *One of the respected elders of engineering admonishes young men entering the profession to follow a "policy of involvement." Write an essay on what you suppose he means.*

OPPORTUNITIES AND CHALLENGES

THE engineer has a wide choice of types of work, fields of specialization, and problem areas, which suit a broad range of talents and interests.

Variety

Many Specialties There are the traditional branches of engineering (page 211) and a number of relatively new ones, like those concerned with space travel, medical instrumentation, and information systems. In addition, in practice most engineers concentrate their efforts on one phase of a major specialty. For example, some mechanical engineers are experts in the design of mechanisms, others concentrate on refrigeration, and another group deals in propulsion systems. Thus, considering all the branches of engineering and the numerous subdivisions of each, you can see that the engineer has a wide variety of specialties from which to choose.

Incidentally, you may find that a curriculum to prepare you for the specialty in which you are interested is unavailable at the college you are attending, or indeed unavailable anywhere. This may happen with some of the newly emerging specialties. But this situation need not deter you from entering the field. Engineers often specialize in one branch of engineering in college and eventually work in a different one. Furthermore, at many colleges it is possible to prepare for such a field by supplementing one of the traditional curricula with appropriate elective courses and informal study.

Many Types of Activity Recall the spectrum of problems described on page 161. Typifying one extreme is task A, which demands thorough up-to-date familiarity with science, technology, and mathematics. Much of this work is creative. Engineers who are happy and successful at this type of job thrive on a steady diet of complex technical problems. Representing the other extreme, task B involves a minimum of technical innovation but a maximum of involvement with people. Superior salesmanship, a keen interest in people, and a pleasing personality are highly desirable for this type of activity, often referred to as "sales engineering." Between these extremes there are hundreds of jobs differing in day-to-day activities, challenge offered, and technical demands. They suit a broad range of abilities and preferences. Take your choice. All jobs over this spectrum involve problem solving, all involve working with things and with people, all require salesmanship, and all require creativeness and the other qualities of an engineer outlined in Chapter 4, but in varying degrees.

Many Types of Industries   Among the traditionally heavy employers of engineers are the aircraft, appliance, automobile, chemical, communications, construction, electronics, energy, metals, machinery, and transportation industries. Problems are becoming more technical and the tempo of innovation in products, services, and manufacturing processes is accelerating, so that the demand for engineers in these industries continues to increase. Even the industries that for decades have survived with relatively stable products are being pressured into major programs of innovation so they too are seeking more engineers. Then there are the newer "glamor industries," like those associated with space and computers; they have come on very strong as employers of engineers.

Many Other Opportunities   A majority of engineers are employed by industrial firms, public utilities, consulting firms, and contractors, but not all of them. Local, state, and federal governments hire increasing numbers of engineers. Also, many are in business for themselves, often as consultants. Furthermore, an engineering education is excellent preparation for a wide variety of careers in other areas. For example, a majority of the executives in many companies have had engineering educations. Many engineering graduates become salesmen, some enter teaching, some become scientists, and so on. Also, persons educated in two professions, such as engineering and law or engineering and medicine, are highly valued.

Thus an engineering education is a valuable background for many careers, both technical and nontechnical. This is understandable. A sharp and well-disciplined mind is a major asset to a person in almost any field of endeavor.

A Challenge

The following pages provide a sampling of fields offering excellent opportunities for engineers over the next several decades. In general, the examples selected are emerging problem areas not as widely publicized as, say, space exploration, yet offering unusual opportunities for employment, advancement, and expression of your social concern. These fields are worth considering *now* because they could affect the program of specialization you select in college.

Storing and Disseminating Information   There are those who claim that mankind's accumulation of general knowledge is now doubling approximately every ten years. Although this claim is difficult to substantiate, it is true that new knowledge is accumulating at a phenomenal rate. Society is truly experiencing an "information explosion." Storage of all this knowledge so that it is reasonably accessible is a challenging problem indeed. We should have effective information storage-retrieval systems from which problem solvers in engineering, business, government, medicine, etc., can quickly, economically, and comprehensively learn what is known about a given problem. The benefits will be superior solutions, better utilization of information, and a significant reduction in costly duplication of effort.

Communications   Although our communications systems are highly developed, their capacities must be constantly extended. The communications industry is always seeking new means of increasing the message-carrying capacity of its systems. One result of such efforts is the fascinating device that enables a single wire to carry simultaneously a large number of telephone conversations as if all were being transmitted over individual circuits. For example, one wire of a modern transoceanic submarine cable handles

128 telephone conversations, all mixed together as they travel along the ocean floor and neatly untangled when they reach the other continent. This sharing principle, which greatly extends the capacity of a communications channel, has wide applicability in the field of communications.

This industry also searches for innovations that provide new communication media. One result of such efforts is the communications satellite, exemplified by Early Bird (now referred to as Intellsat I), the sensational development for relaying telephone conversations and live television between distant parts of the world. Earlier communications satellites like Telstar moved rather rapidly with respect to the earth and therefore were simultaneously "in view" of the transmitting and receiving stations for relatively short periods of time. Continuous transmission could only be achieved by providing dozens of such satellites, so that at least one was always in view of both stations. But Intellsat I does not move with respect to the earth. It is orbiting at a speed which matches that of the earth's surface, so that it is permanently "parked" 22,300 miles above a fixed point on the earth. It is referred to as a synchronous satellite, since its movement in orbit is synchronized with the rotation of the earth. Three synchronous satellites stationed as shown in Figure 1 can relay communications between most of the populated areas of the globe—an intriguing and

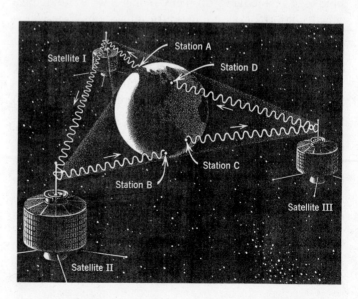

Figure 1   Three synchronous communications satellites positioned as shown can receive, amplify, and retransmit telephone conversations and TV transmissions between most areas of the world. For example, satellite III is always simultaneously in view of stations C and D. Satellites I and II can relay signals between stations A and B, which are on opposite sides of the earth.

elegant concept indeed. Ideas like this, plus the apparently insatiable demand for more communication channels, plus the prospect of a large and prosperous satellite-communications industry, plus the fact that expansion and innovation have long been conspicuous characteristics of the communications industry, are assurances that this field will be lively, challenging, and rewarding for engineers for a long time to come.

Transportation   There are three major deficiencies in our transportation system: disproportionately slow interconnections, inbalance, and neglect of Everyman. These weaknesses are worth exploring further because engineers can contribute much to remedy all three.

Picture the traveler who covers 2450 miles of his 2500-mile journey in four hours, and then spends four hours between the door of the plane and the door of his home 50 miles away. One reason for this gross inconsistency is the typically slow time to transfer from one medium to another, for example, plane to bus and then bus to train. The situation is a familiar one, and it is no different for the transportation of goods.

Another deficiency is the present inbalance in our travel capability over different distances. Today the person who wants to go 50 or 100 miles or so doesn't have it nearly so good as the one who is making a 400-mile trip. Especially in the more densely populated areas is this inbalance evident.

A third significant weakness is what many people feel is a neglect of Everyman—and I concur. The quotation on page 175 sums this up very nicely: "... Everyman ought to be about due for a break." We need faster, cheaper, less frustrating *mass* transportation, *over the whole range of trip distances,* intraurban through internation. At present, if you are traveling more than 100 miles and you have the money, you're all set: you go jet; otherwise . . . .

There are many exciting transportation ideas in various stages of development; you can get some idea of these possibilities from Figures 2–11. Besides those shown, there are opportunities in such areas as battery-powered automobiles, VTOL's, pipelines for transporting solids, and unmanned surface and subsurface ships for transoceanic transportation of goods. Incidentally, the reason that I have not included numerous illustrations of "ideas in evo-

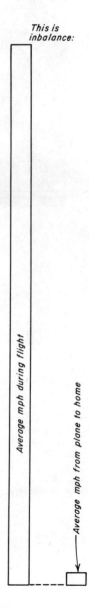

This is inbalance:

Average mph during flight

Average mph from plane to home

(a)

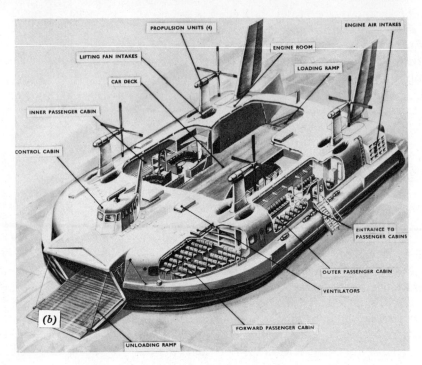

(b)

LIFTING FAN INTAKES

PROPULSION UNITS (4)

ENGINE ROOM

ENGINE AIR INTAKES

CAR DECK

LOADING RAMP

INNER PASSENGER CABIN

CONTROL CABIN

ENTRANCE TO
PASSENGER CABINS

OUTER PASSENGER CABIN

VENTILATORS

FORWARD PASSENGER CABIN

UNLOADING RAMP

(c)

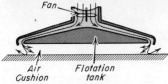

Fan

Air
Cushion

Flotation
tank

Figure 2 These are commonly referred to as air cushion vehicles (ACV). An ACV is supported by a cushion of air trapped within its flexible skirts, enabling it to travel over land and water with ease. The vehicle shown in (a) and (b) carries 600 passengers, cruises at 65 mph, even over 10 foot waves. Craft of this type provide ferry service across the English Channel. (Courtesy of British Hovercraft Corporation.) The proposed ACV freighter pictured in (c) would leave any conventional ship in its spray, cruising at 90 mph. This 420-foot, 4000-ton craft portrays what should be a common transporter of people, goods, and military equipment in the future. (Courtesy of Textron's Bell Aerosystems Company.)

lution" for each problem area in this chapter is simply lack of space; exciting developments are "in process" in all of them.

The opportunities for you in a field like transportation are of three types: conceiving new systems; developing ideas like those pictured here into economical, reliable, safe, and otherwise satisfactory systems for large-scale use; and helping to solve the nontechnical problems that currently are impeding the effective introduction of new transportation systems. This last area of opportunity requires some elaboration.

From Figures 2–11 it is obvious that we are not hurting for ideas, yet we are finding it difficult to get these beyond the concept or demonstration stage. Why? Well, here are a few of the impediments: special-interest pressure groups, political boundaries that inhibit interregional co-operation, the "narrow-sightedness" of problem solvers and policy makers, outmoded laws, overlapping spheres of interest of government agencies, biases, and local conflicts.

Figure 3 Even if the use of VTOL's becomes widespread, there will still be critical problems in locating and designing airports. Through some imaginative and unconventional thinking like that demonstrated here and in Figure 4, these problems may be solved. This is a proposed jetport over water, a scheme which might well be the only answer for some areas. (Courtesy of Los Angeles Department of Airports.)

Figure 4 A promising proposal for reducing door-to-door time for air travelers is the detachable passenger pod. It can be boarded at strategic locations around the city and then picked up and taken to the plane. In fact, we may someday find it unnecessary to transfer from pod to plane. (Courtesy of Los Angeles Department of Airports.)

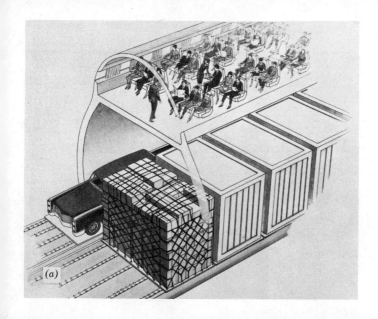

(a)

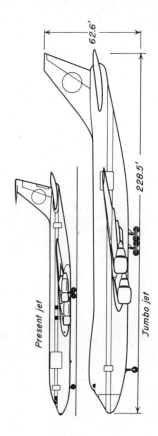

62.6'

228.5'

Present jet

Jumbo jet

(b)

Figure 5 Jumbo jets bring air travel within the means of a much larger proportion of our population. Planes this size carry about 500 persons and cruise at more than 600 mph. View (a) is a cross section of the passenger-cargo version. From (b) you can get some idea of the cabin size of the "first generation" jumbo jets. (Courtesy of the Lockheed-Georgia Company, Wide World Photos.)

187

Figure 6 *New schemes for bus terminals like that proposed here, plus bus-trains and other ways of expediting the flow of buses, are promising ideas. (Courtesy of General Motors Corporation.)*

Figure 7 *Many new high-speed train concepts are under consideration. One is the air-supported train shown here. (Courtesy of Bertin et Compagnie and Societé de l'AEROTRAIN.)*

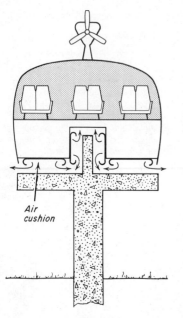

Air
cushion

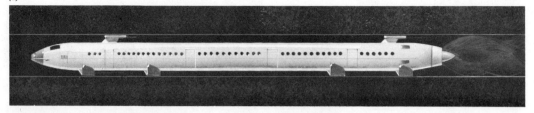

(a)

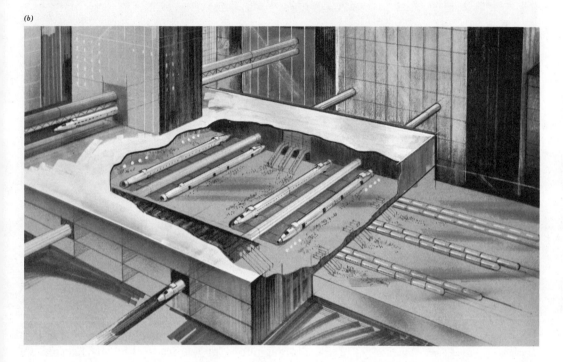

(b)

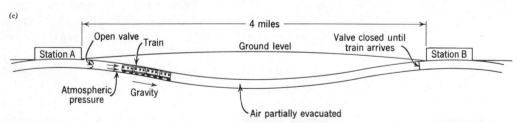

(c)

Figure **8**   *Another high-speed train concept being explored is the tube scheme. It is especially well suited for densely populated areas like the Northeast because the tube can carry trains underground. (a) shows one of the designs considered; it literally flies through the tube, supported on a cushion of air. It takes air in at the front and discharges it under pressure at the rear. (b) pictures a tube-train terminal of the future. (Courtesy of Project Tubeflight.) (c) diagrams the intriguing gravity-vacuum concept introduced by L. K. Edwards, engineer and president of Tube Transit Corporation. When the train leaves station A it is accelerated by gravity and an air pressure differential. When it approaches the next station a pressure buildup in front of it, and gravity, cause it to decelerate and come to a stop in the terminal.*

Figure 9   Here is a forerunner of a new generation of high-speed urban transit systems. You see one portion of the new one-billion-dollar BART system (Bay Area Rapid Transit). Trains are computer controlled and can run at 70 mph at 90-second intervals. About 4 miles of this system passes beneath the bay in tubes very much like those used in the Chesapeake Bay bridge tunnel. (Courtesy of San Francisco Bay Area Rapid Transit District.)

Note that these obstacles are social, political, and legal, not technical.

This pattern is certainly not limited to the field of transportation. The major obstacles to progress in many problem areas are shifting from the technical to the human. In many instances we have the technical capability to go ahead, but implementation is impeded by complex social, political, and legal knots, as well as ignorance on the part of the general public of what is technically feasible. Of course all this should remind you of remarks in the preceding chapter concerning the need for more engineers to participate in public affairs. The call is for *involvement* and *action* by engineers.

I am convinced that whatever engineers can do, through improved communication and transportation for the masses, to aid peoples of different nations to know one another better will significantly improve the chances for peace. And similarly for the movement of goods—lowering the cost of transportation between nations increases foreign trade, which increases interdependency between nations, which in turn increases what each stands to lose through animosity and conflict. See what you can do about it!

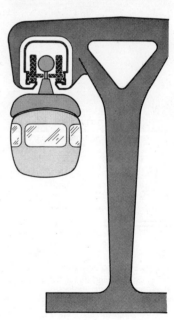

Figure 10 Here are two of a number of overhead systems that have been proposed. These are especially well suited for medium-sized cities because these operatorless cars can be run individually and in trains, at high speed and short intervals under computer control. (Courtesy of Westinghouse Electric Corporation and General Electric Company.)

Under driver control

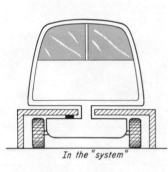

In the "system"

Figure 11 *This system has a number
of advantages offered by private
automobiles and high-speed mass
transit systems. (a) This man starts out
in his electric auto, drives about the
suburbs (b), and when ready to head
for center city, he drives to a high
speed track which whisks him there (c).
The track provides power and controls
the speed, spacing, and discharge of
the vehicles. At his destination he
leaves the vehicle which is stored or
used by someone else. When he
is ready to return home in the evening,
a vehicle (probably not the one he used
that morning) is at the station
awaiting him. (Courtesy af Alden
Self-Transit Systems Corporation.)*

Education Many of us are unhappy with our current educational system, in particular with the uniform pace that students must keep as members of a large group of identically classified pupils. Teaching machines, especially the computer-tutor, offer promising means of overcoming this major weakness. The computer-tutor is a time-shared computer that carries on a dialog with, say, 50 students simultaneously and independently (in fact, they can be taking different courses). Under this system information is presented to each student individually by means of conventional texts, teletypewriters, earphones, or graphic displays (page 93). Frequently the computer asks questions and the student responds via the typewriter or light pen. If the student does not understand the information, the computer synthesizes remedial material and presents it immediately. This cycle of events is repeated until the student learns the lesson. Thus each student progresses as rapidly as his ability and inclination dictate. Under this system the level of achievement tends to be the same for all students and the time required to reach it is allowed to vary; in the conventional system the opposite is true. Although the educational principles employed here are those of the psychologist, the equipment is engineering's contribution. These systems are becoming quite sophisticated and promise to be even more so in the future. Other innovations, such as new means of communication between student and machine, are also in the offing in this rapidly expanding field.

Engineers can contribute in other ways to the educational process. Especially challenging and beneficial is the improvement of education in underdeveloped areas of the world by developing low-cost structures that are easily constructed in remote areas, power supplies for schools where there are no power lines, and devices that aid learning and help to compensate for the shortage of skilled teachers.

Engineering's Potential Contribution to Medicine The modern operating room is filled with impressive-looking machines that are now routinely employed. Then too there are such engineering accomplishments as the heart-lung machine (Figure 12), the miniature devices for incisionless internal surgery, and the equipment for neurosurgery by freezing.

For medical diagnosis there are tiny instruments that

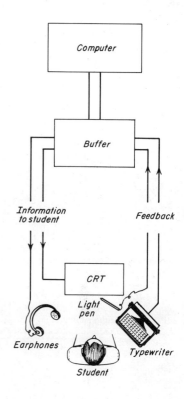

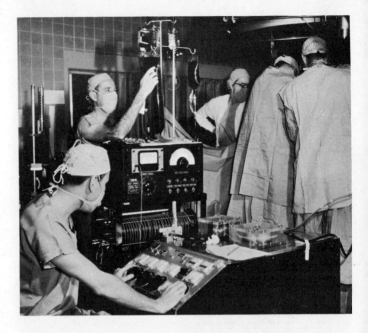

Figure 12   The heart-lung machine fulfills the functions of heart and lungs during operations on these organs. Many patients owe their lives to the designers of this machine. (National Institutes of Health photograph.)

can be inserted within vital organs (including the heart via a vein) to make measurements; apparatus that automatically analyzes blood samples, and computerized analysis of electrocardiagrams. Other diagnostic innovations are in the development stage.

To aid in medical treatment there is the miniature battery-operated heart pacer which is embedded in the patient's body to keep his heart operating by sending regularly spaced electrical pulses directly into the heart muscles. A number of artificial organs and vastly improved artificial limbs are under development. One of the most dramatic is the artificial kidney.

In hospitals some exciting developments are taking place. A variety of ingenious sensors are being developed for continuous measurement of patients' temperature, blood pressure, etc., so that these characteristics can be monitored at a central station. This monitoring system sounds an alarm if a change in a patient's condition warrants attention. The tiny cardiac sensor illustrated in Figure 13 indicates what is to come in the field of patient care.

In spite of these developments major difficulties remain, such as the information problems of hospitals and the ever-present matter of making new "hardware" widely available at reasonable cost. So there are excellent oppor-

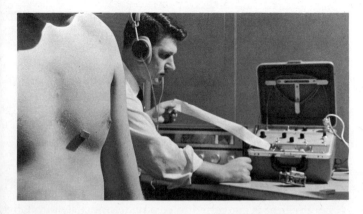

Figure **13** *This tiny cardiac monitor affixed to the man by adhesive tape contains a sensor, an amplifier, a radio transmitter, and its own power supply. It senses the action of the heart and transmits the information to the recorder shown in the background. (Courtesy of the United Aircraft Corporation.)*

tunities for those who wish to apply their engineering talents in the medical field. Because the potential is so great, a new branch of engineering, called biomedical engineering, is emerging. This specialty is concerned primarily with the development of instrumentation for research, diagnosis, treatment, prosthesis, and patient care and rehabilitation.*

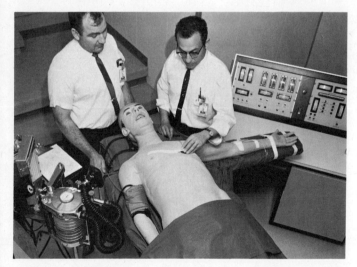

Figure **14** *This computer controlled simulator, developed to train anesthesiologists, is shown being prepared for demonstration by two of the engineers primarily responsible for its design. This patient "breathes, coughs, and vomits," has a "heart beat," and can "die." In these and many other respects Sim One closely mimics human behavior. Students can give injections, administer oxygen, check pulse rate and pupil dilation, and perform most other routine and emergency functions of an anesthesiologist. The instructor, using the console at right, can program the "lesson," evaluate performance, temporarily stop the process to discuss a point with the student, and have a lesson repeated if performance is not satisfactory. Obviously, these are privileges not available when real subjects are used. This simulator is welcomed especially because it significantly shortens the training period for an anesthesiologist and allows instructors to exert much closer control over the learning process. (Courtesy of the Aerojet-General Corporation.)*

Natural Resources   The supplies of oil and natural gas are dwindling, whereas the demands are increasing. Efforts to develop economical energy resources to replace them

* Educational and career information can be obtained from Instrument Engineering and Development Branch, Division of Research Services, National Institutes of Health, Bethesda, Maryland.

are beginning. Here is another problem area that will emerge into major significance in the future. Elaborate efforts are already directed to the development of economical means of converting nuclear to electrical energy. This is emerging as a big field in itself. However, a number of other sources of electric power have potential commercial applications. Among the sources currently under development are solar energy, the fuel cell, the thermoelectric generator and the thermionic converter (both of which convert heat to electricity by a process that is relatively direct in contrast to the familiar heat-steam-turbine-generator system), and the magnetohydrodynamic generator (whereby heat is converted to electricity primarily through the medium of superheated ionized gases). The big task in each of these cases is to develop an economical means of generating electric power on a large scale by the process in question.

Another resource that needs bolstering in many parts of the world is fresh water. By 1980 the consumption of fresh water will be almost double what it is now. A water crisis will have developed by the year 2000 if new sources of major proportions are not developed. As the population explosion continues, it will soon be necessary to occupy the presently arid areas of the earth. Large supplies of fresh water are needed to prepare these lands for vegetation and habitation. One indication of the current interest in this problem are the extensive efforts to devise processes for economical conversion of large quantities of sea water to drinkable form. In the future, engineering will be devoting more of its efforts to fresh-water sources and conservation, including measures to reduce pollution of streams and lakes.

Oceanography and Aquaculture   Recently there has been a surge of interest in oceanography. One result is a flurry of specially designed research craft like those shown in Figures 15 and 16. Much engineering will continue to go into the development of machines and instruments for ocean exploration.

Much engineering activity will also be devoted to means of extracting the ocean's great resources. Man's needs for minerals, oil, food, and fresh water are rapidly increasing, while the land sources of these resources are diminishing. One area of activity, aquaculture, is concerned with the

(a)

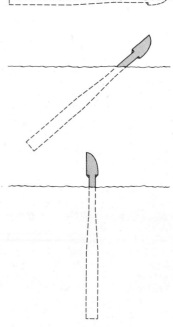

(b)

Figure 15 Persons in nearby boats
are startled, to say the least, when
the bow of the vessel that they are
watching begins to rise out of the
water and the stern to sink, until
the vessel is literally floating on end,
as pictured here. This is a floating
laboratory and home for oceanographic
researchers conducting studies of
underwater acoustics, marine biology,
and other matters associated with
the oceans. In this position, approximately 300 of its relatively thin,
350-foot hull is below water. It is
remarkably stable in this position,
relatively unaffected by waves and
swells. (Courtesy of Scripps Institution
of Oceanography.)

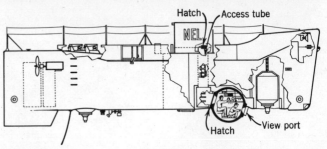

Figure 16   The number and variety of undersea vehicles in use or under development would make another interesting case study of engineering alternatives. This is Trieste II, one of the twenty plus manned submersibles in existence. It can reach 20,000 feet, the deepest so far, with a crew of 3 men. (Official U. S. Navy Photograph.)

cultivation and harvesting of plant and fish food. Some intriguing underwater structures and ingenious machines will evolve for this purpose. Another opportunity area is the development of means of economically extracting minerals from the ocean and beneath it. The ocean is already providing substantial amounts of oil and sulfur by way of colossal-sized off-shore facilities (Figure 17, pp. 200–201). Such industries are bound to grow rapidly.

Urban Problems   As a result of the rapid growth of urban areas, of industrialization, of the unplanned manner in which cities have evolved, of the rapid increase in the automotive population, and a number of other factors, an aggregation of new and interrelated problems faces the urban community. Some of these are congestion, inadequate transportation media for masses of people, inadequate housing, air and water pollution, diminishing water supplies, rising noise levels, unsatisfactory distribution systems for utilities, snow removal, and vehicle parking. There is special need for engineers who are concerned about the *total* urban-improvement problem.

Travel Safety   The creations of engineers have provided sensational improvements in some modes of travel. However, improvements in safety have not kept pace with those in speed, comfort, capacity, and cost. Furthermore, as the numbers of surface and air vehicles continue to increase at accelerating rates, safety of the roadways and that of the airways become pressing problems. An engineer can find excellent opportunities in the fields of highway and airway safety.

There are some promising possibilities for improving highway safety through better vehicle and roadway design and through vehicle guidance systems. For example, shock-absorbing auto frames and shells, fast-inflating balloons that fill on impact and prevent passengers from leaving their seats, and many other possibilities may significantly improve auto safety. Highway guidance systems are in the early stages of development (Figure 18, p. 202).

There is need for engineers with positive attitudes toward auto and highway safety possibilities. I do not accept the assumption that a safer automobile must cost more. I am convinced that we can design more "foolproof" features into highways and interchanges. But open minds and imaginative thinking are needed.

Then there are the airways, where the traffic problem is becoming critical. Needed: an improved system for controlling in-flight traffic, a system that will take over from the pilot and land the plane safely under any atmospheric conditions, a collision-avoidance system, and improved terminal facilities for the densely populated sections of the world. Society will be most grateful to the engineers who satisfactorily solve these problems.

Underdeveloped Nations   The solution of many of the world's economic, political, and social problems depends heavily on the solution of technical problems. In areas of the world classified as underdeveloped there are some crucial needs: vastly increased food-production capacity, facilities for extraction of natural resources, sources of electric power, improved transportation systems, education of the masses, *and* training of the inhabitants to cope

Figure **17a** *An oil well-drilling platform which stands on retractable
legs while doing its work. When it must
be moved, the legs are pulled up and the
platform is towed to the next site. It
can stand in water up to 300 feet
deep, and if it chose to "walk"
downtown it could straddle a 30-story
building. (Courtesy of R. G.
LeTourneau, Inc.)*

with their technical problems. For some time to come there
will be a pressing need in these nations for outside engineering assistance. The sense of accomplishment associated with this type of endeavor and the concurrent experi-
ences are rewarding indeed. As an engineer you can
contribute in this respect through a variety of means (Figure 19, p. 203).

Conclusion An analysis of the problem areas in which
engineering stands to make significant contributions re-
veals a special need for engineers who bring to bear on
problems a *broader perspective, more imagination,* and *a
higher sense of purpose* than have prevailed up to now.

On this matter of a broader perspective, interrelated
problems, such as the various ones plaguing the urban

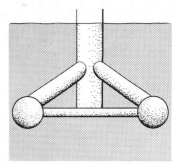

*Figure **17b** Another of the many types of ocean drilling platforms in use. This one is towed to the site, which in this instance is the turbulent North Sea, where it is partially submerged and anchored. This rig houses 50 men and has a platform area of 30,000 square feet. (Courtesy of the British Petroleum Company.)*

*Figure **17c** This "monopod" drilling platform is shown on duty off Alaska. It too is towed from site to site and partially submerged while drilling for oil. This is a larger structure than you might conclude from this photograph. Remember, much of it is beneath the surface. (Courtesy of the Union Oil Company of California.)*

201

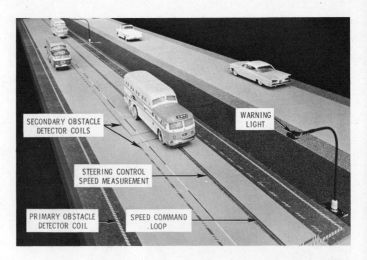

Figure **18** *This is a working model of an "automatic highway" that guides vehicles, maintains their speed and spacing at preset values, and applies their brakes when obstacles are detected. Vehicles equipped with an electronic device will follow a wire embedded in the inside lane when the driver switches to automatic control. The outside lane is used by manually controlled vehicles. This system is under development. (Courtesy of the General Motors Corporation.)*

community, are too often attacked separately and independently. Often a solution in one area has aggravated the situation in another. Foresight has been lacking, so that, for example, we find sprawling residential areas where no provision has been made for highways and airports. Engineers are needed who will view a situation with broader perspective than has been customary and produce a well-integrated solution to the whole problem.

There is need also for engineers who will bring fresh ideas to the many long-standing, still unsatisfactorily solved problems. Society can use engineers who are bent on *imaginative* application of their knowledge and skills.

Engineering needs many more practitioners dedicated to applying their unique qualities where society's needs are greatest. Automobiles of greater horsepower are hardly one of mankind's pressing needs. There *are* many challenging, fascinating problem areas in which you can make significant contributions. I have mentioned a few of them on succeeding pages. You *can* do something to alleviate conditions that breed social ferment, political upheaval, and conflict between nations, to improve and extend educational facilities, to aid the physically disabled, to better the urban environment, to improve the safety, cost, and availability of modes of travel, to combat starvation and disaster, to deter and detect crime, or to convert the great wastelands into useful forms. These are a few of the outlets engineering offers for your social concern.

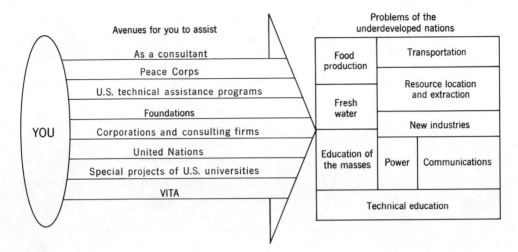

Exercises

1 *Prepare a paper entitled "Engineering Problems Created by Engineers." (Air pollution, traffic congestion, and intercontinental ballistic missiles are examples.)*
2 *Prepare a paper entitled "Engineering in 1985."*
3 *Write a paper summarizing engineering's past and potential contributions in one of these fields:*
   (a) *Surgery.*
   (b) *Water-resources development.*
   (c) *Prosthetics.*
   (d) *Air safety.*
   (e) *Oceanography and utilization of the sea's resources.*
   (f) *Hospital design and operation.*
   (g) *Urban environmental control.*
   (h) *Air cushion vehicles.*
   (i) *System reliability.*
   (j) *Use of electronic computers in design.*
   (k) *Space research.*
   (l) *Lasers.*
   (m) *Cryogenics.*
   (n) *Biological research.*
4 *It would be helpful to your career planning to have an alternatives tree (page 139) for the many possibilities in and closely related to engineering. Continue with the "tree" begun in Figure 20. Some of the suggested readings should be consulted. (Don't try to fit this tree on an 8½ by 11 inch sheet of paper!)*

*Figure **19** There are many ways you can use your engineering competence to aid the underprivileged people of the world (VITA stands for Volunteers for International Technical Assistance, College Campus, Schenectady, New York 12308.)*

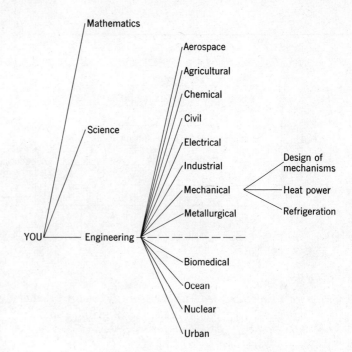

Figure 20 Some of the alternatives a person interested in engineering should consider. For each specialty there are a number of subspecialties, as illustrated here for mechanical engineering.

SUGGESTED READINGS

Chapter 2

Goode, Harry H., and Robert E. Machol, *Systems Engineering*, McGraw-Hill, New York, 1957. Chapter 2 describes a number of interesting engineering projects, all involving rather large systems.

O'Brien, Robert, and the editors of LIFE, *Machines*, TIME, Inc., New York, 1964. Many engineering achievements are described, mainly through drawings and photographs.

Rapport and Wright, eds., *Engineering*, New York University Press, New York, 1963. Detailed descriptions of the background and evolution of a number of well-known engineering achievements are interestingly presented.

Chapter 3

De Camp, L. Sprage, *The Ancient Engineers*, Doubleday, Garden City, N.Y., 1963. The evolution and accomplishments of the early engineers are described, including the Egyptian, Oriental, Greek, Roman, and other European engineers of old.

Finch, James K., *The Story of Engineering*, Doubleday, Garden City, N.Y., 1960. A good history of engineering with illustrations. Available in paperback (Anchor 214) for $1.45.

Furnas, McCarthy, and the editors of LIFE, *The Engineer*, TIME, Inc., New York, 1966. Beautifully illustrated description of engineering, past and present.

Kirby, Richard S., et al., *Engineering in History*, McGraw-Hill, New York, 1956. An interesting, well-illustrated history of engineering, covering all branches of the profession.

O'Brien, Robert, and the editors of LIFE, *Machines*, TIME, Inc., New York, 1964. The creative nature of engineering is dramatically illustrated in this interesting description of machines invented over the centuries.

Sporn, Philip, *Foundations of Engineering,* Macmillan, New York, 1964. Short, interesting, written by a man with a wealth of engineering experience.

Usher, Abbott P., *A History of Mechanical Inventions,* rev. ed., Harvard University Press, Cambridge, 1954. A scholarly discussion of inventors and inventions, from early times until the last century.

Yost, Edna, *Modern American Engineers,* Lippincott, Philadelphia, 1958. The lives and achievements of twelve well-known contemporary engineers.

Chapter 4

A Professional Guide for Young Engineers, Engineers' Council for Professional Development, 345 East 47 Street, New York. Price $1.

Johnson, Lee H., *Engineering: Principles and Problems,* McGraw-Hill, New York, 1960. Chapters 4–7 are introductions to several important engineering skills, including measurement and the use of mathematics.

Beakley and Leach, *Engineering: An Introduction to a Creative Profession,* Macmillan, New York, 1967. Chapters 8–16 elaborate on some of the skills not treated in detail here.

Chapter 5

Bross, Irwin D. F., *Design for Decision,* Macmillan, New York, 1953. There is an excellent introductory discussion of models in Chapter 10.

Walker, Marshall, *The Nature of Scientific Thought,* Prentice-Hall, Englewood Cliffs, N.J., 1963. The philosophical discussion of models, their significance and their development, presented in Chapter 1 is the best I have seen.

Woodson, Thomas T., *Introduction to Engineering Design,* McGraw-Hill, New York, 1966. Rather extensive treatment of various types of models in Chapters 10–12.

Chapter 6

Woodson, Thomas T., *Introduction to Engineering Design,* McGraw-Hill, New York, 1966. See Chapters 13 and 15.

Wilson, Warren E., *Concepts of Engineering Systems Designs,* McGraw-Hill, New York, 1965. Pages 127–138 on optimization are good.

Chapter 7

Burck, Gilbert, *The Computer Age,* Harper and Row, New York, 1965. A collection of excellent articles from FORTUNE.

Desmonde, William H., *Computers and Their Uses,* Prentice-Hall, Englewood Cliffs, N.J., 1964.

Fetter, William A., *Computer Graphics in Communication,* McGraw-Hill, New York, 1966. A good survey of the nature and usefulness of computer graphics.

Information, a SCIENTIFIC AMERICAN book, W. H. Freeman, San Francisco, 1966. Probably the most comprehensive and comprehensible book on computer applications. The price of $2.50 yields a high benefit-cost ratio.

Thornhill, Robert B., *Engineering Graphics and Numerical Control,* McGraw-Hill, New York, 1967. Rather practical and detailed treatment of computer graphics.

Also, computer companies have brochures and manuals describing computer applications in engineering. Write and ask for these publications.

Chapter 9

Asimow, Morris, *Introduction to Design,* Prentice-Hall, Englewood Cliffs, N.J., 1962. Chapters 7 and 8, on the design process and problem analysis, are especially worth reading.

Chapter 10

Alger and Hays, *Creative Synthesis in Design,* Prentice-Hall, Englewood Cliffs, N.J., 1964. Written specifically for engineers.

Dixon, John R., *Design Engineering: Inventiveness, Analysis, and Decision Making,* McGraw-Hill, New York, 1966. Chapter 2 is excellent.

Osborn, Alex F., *Applied Imagination,* Scribner, New York, 1957. A lively discourse on creativity, written from a general point of view.

Von Fange, Eugene, *Professional Creativity,* Prentice-Hall, Englewood Cliffs, N.J., 1959. A practical book with many excellent suggestions for improving your inventiveness.

Whiting, Charles S., *Creative Thinking,* Reinhold, New York, 1958. Strong on brainstorming.

Chapter 11

Grant, Eugene L., and W. Grant Ireson, *Principles of Engineering Economy,* 4th ed., Ronald Press, New York, 1960. A sound introduction to the economics of engineering.

Starr, Kenneth S., *Product Design and Decision Theory,* Prentice-Hall, Englewood Cliffs, N.J., 1963. The first three chapters describe the general decision-making process in design, and the means of taking into account the differing degrees of uncertainty ordinarily associated with alternative solutions.

Chapter 12

Asimow, Morris, *Introduction to Design,* Prentice-Hall, Englewood Cliffs, N.J., 1962. Although not an introductory book, this is a good intermediate-level treatment of general design methodology.

Buhl, Harold R., *Creative Engineering Design,* Iowa State University Press, Ames, Iowa, 1960. An introduction to design with emphasis on creativity.

Chapter 14

Citizenship and Participation in Public Affairs, Engineers' Council for Professional Development, 345 East 47th Street, New York. Price 20 cents.

Davenport and Rosenthal, *Engineering: Its Role and Function in Human Society,* Pergamon, New York, 1967. An excellent anthology of writings on the role and impact of engineering, by prominent persons in many fields.

Lilley, S., *Man, Machines, and History,* Cobbett, London, 1948. A short history of tools and machines in relation to social progress, well illustrated and interestingly presented.

Sporn, Philip, *Foundations of Engineering,* Macmillan, New York, 1964. An inspiring little book by a well-known and highly respected engineer.

Walker, Charles R., *Modern Technology and Civilization,* McGraw-Hill, New York, 1962. A scholarly collection of writings by a number of authorities on the impact and problems created by engineering creations.

Chapter 15

Love and Childers, eds., *Listen to Leaders in Engineering,* David McKay, New York, 1965. Extensive description of opportunities in many branches of engineering, written by authorities in these fields.

O'Neill, John J., *Engineering the New Age,* Ives Washburn, New York, 1949. An inspiring book on the opportunities awaiting future engineers in the advancement of civilization and the betterment of mankind.

Whinnery, John R., *The World of Engineering,* McGraw-Hill, New York, 1965. Good description of opportunities for engineers in both the traditional and especially the evolving specialties.

MAJOR BRANCHES OF ENGINEERING

1   These are the major, not the only, branches of engineering—there are hundreds of subspecialties within this field.

2   Very few modern engineering creations can be designed completely by a man educated in any one of these branches. Design of a power-generating station involves civil, electrical, and mechanical engineering problems. *Problems of the real world seldom follow the same boundaries that exist between branches of engineering in college.*

3   Education in a particular branch does not forever bind you to it in practice. Far from it! In fact a substantial percentage of graduates end up in a specialty different from their major in college. An engineering education is excellent preparation for a wide variety of jobs in and out of engineering, regardless of what branch of the field you study. The same basic characteristics are identifiable in every branch: design of physical systems to transform material, energy, human, or information resources into useful forms, calling for the *same basic skills,* the *same point of view,* and even much of the *same basic knowledge.*

Major Branches

Aeronautical   Design primarily of aircraft and associated systems. Examples: all types of conventional airplanes, vertical-take-off-and-landing craft (VTOL), landing systems, collision avoidance systems, air cushion vehicles. (Two related specialties are evolving. Astronautical engineering is directed to flight beyond the atmosphere. Aerospace engineering is concerned with all travel above the earth's surface.)

Chemical   Design primarily of processes for the *chemical* transformation of materials. Examples: facilities for the production of gasoline, paint, explosives, rubber, cement.

Civil   Design primarily of major structures *and* the means of constructing them. Examples: highways, bridges, dams, canals, water supply and sewage disposal systems, airports, and waterports.

Electrical   Design primarily of the means by which electrical energy is created, transferred, and used. Examples: electrical generators, transmission networks, communication systems.

Industrial   Design primarily of systems for the *physical* transformation of materials. Examples: automobile plant, printery, shipyard, textile mill.

Mechanical   Design primarily of the systems by which energy is converted to useful mechanical forms. Examples: turbines, motors, *and* the mechanisms required to convert the output of these machines to the desired form, including pumps, compressors, and transmission systems.

Metallurgical   Design primarily of means of extracting, processing, and fabricating metals. Examples: aluminum refinery, mill for rolling metal into flat sheets, equipment for extruding metal, processes for making and shaping powdered metal, equipment for testing and inspection.

The headquarters for most of the engineering professional societies are in the United Engineering Center, 345 East 47 Street, New York, New York, 10017. If you wish more information about a particular branch of engineering, write there and ask for it.

DIGITAL SIMULATION BY COMPUTER

To convert the parking lot simulation (page 60) to a form executable by computer we must find a way for the machine to carry out the Monte Carlo procedure. The spinner and the paper-slips methods obviously will not do here. But this conversion is not difficult.

To understand how it is done, assume you have a spinner device (margin) which has equally spaced marks numbered 1–100 around its circumference. You spin the pointer and record its stopping point in terms of this 1–100 scale. The results of many repetitions of this procedure are shown in Table 1. (Note that in the long run each of the numbers 1

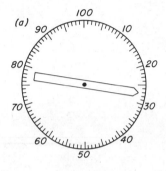

Table 1

Table of Random Numbers				
12	82	89	54	11
74	41	21	02	13
81	06	19	79	91
19	43	17	75	82
29	21	35	18	57
47	98	81	96	28
26	60	24	77	49
17	21	53	91	38

through 100 will appear with equal frequency in such a table, but in random order.) With this table of random numbers available, the pointer of the spinner device can be discarded. Now you merely read a number from Table 1 and consult pie chart (b) to determine whether this number falls in the no-show sector. Obviously with this table you don't need any part of the spinner device. Simply select a number N from the table and apply the following rules. If $N \le 08$, the driver in question is a no show that day. If $N > 08$,

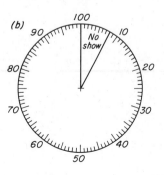

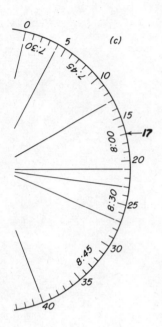

proceed to generate the driver's arrival time. Of course, if you are familiar with computer programming you probably aleady recognize the ease with which the machine can follow these rules and have anticipated the steps diagrammed in Figure 1a.

A similar procedure can be employed to replace the spinner method of randomly selecting a driver arrival time. The pointer of the spinner is replaced by a scale of 1–100 on its circumference, as in (c). Now you can select a number from the table of random numbers, consult this pie chart to learn in which sector the number falls, and assign the driver an arrival time accordingly. For example, if the value selected from the table is 17, it falls in the 8:00 A.M. sector.

Of course, in place of the pie chart you would use these rules:

If N is between 01 and 04 inclusive, assign an arrival
time of 7:30 A.M.;
If N is between 05 and 13 inclusive, assign 7:45 A.M.;
If N is between 14 and 21 inclusive, assign 8:00 A.M.;
.

.

.

If N is 99 or 100, assign 4:00 P.M.

The computer's way of executing this procedure is diagrammed in Figure 1b. In terms of the pie chart the computer is in effect starting with the 7:30 sector and asking, "Random number, do you fall in this sector?" If not, it goes to the next sector and inquires, "Random number, do you fall in this sector?" If not, it continues to step from sector to sector until it determines what arrival time to assign.

If you understand the procedure just described, and if you are able to program a computer, you are equipped to write computer simulation programs for a wide range of engineering problems. This is no small matter, for simulation by digital computer is a very powerful technique of rapidly increasing importance in engineering practice.

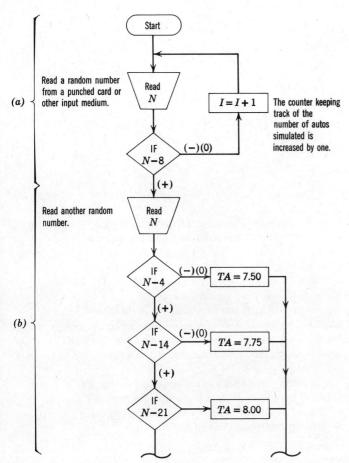

Figure **1** *Diagram of a portion of a computer program for executing the parking-lot simulation described on page 60. If so desired, the computer can generate a random number by means of a special subroutine. TA replaces the previously used T_a. Of course decimal hours must be used in place of minutes, e.g., 7.75 = 7:45.*

Appendix C

A MORE RIGOROUS ANALYSIS
OF A PROBLEM

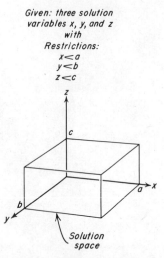

Given: three solution
variables x, y, and z
with
Restrictions:
$x < a$
$y < b$
$z < c$

Solution
space

*Figure 1   Problem with three solution
variables with restrictions on each,
graphed to show the resulting solution
space.*

Picture a problem with three solution variables, x, y, and z. Also suppose that these solution variables are restricted, so that x may not be greater than a, y may not be greater than b, and z may not be greater than c. These limits establish the *solution space* shown in Figure 1. The final solution must come from this space.

In most problems there are more than three solution variables, so the situation is one of N variables in N-dimensional space. When restrictions exist, there are boundaries on this multidimensional space equivalent to those in Figure 1.

A solution variable is an independent variable, and if it is not fixed by a restriction, it is manipulated by the designer to observe its effect on the criterion. Thus, following the pattern established in Chapter 6,

$$C = f(V_1, V_2, V_3, \ldots, V_N),$$

where V_i represents a solution variable.

Recall that in the cleaning-machine problem the cost of the unit could not exceed \$125. For simplicity this was listed as a restriction, but strictly speaking it is a limit on a criterion, not on a solution variable. Any solution for which $C > \$125$ is ruled out. Such a limit on a criterion is called a *criterion constraint*. Thus:

- A *restriction* fixes or limits a solution variable, for example, "the bridge must be at least 75 feet above the water"; "the machine must use 110-volt alternating current."
- A *criterion constraint* limits the acceptable values of a criterion, for example, "the carrying capacity of the missile under design must be at least 2000 pounds."

Be sure you see the difference between these four types of variables in a design problem and their respective limits:

Type of Variable	Limit is Called
Input variable	Input constraint
Output variable	Output constraint
Criterion	Criterion constraint
Solution variable	Restriction

The designer's task is to determine the combination of values for solution variables such that

$$C = f(V_1, V_2, V_3, \ldots, V_N)$$

is a maximum (or minimum, whichever is appropriate), and such that all restrictions and constraints are satisfied.

Since terminology in design literature is not standardized, you may well encounter these terms as alternatives to mine:

- *restriction:* solution constraint, constraint;
- *solution variable:* design variable, independent variable, design parameter, or parameter;
- *criterion:* dependent variable, performance variable, measure of effectiveness.